INTERNATIONAL ASTRONOMICAL UNION

UNION ASTRONOMIQUE INTERNATIONALE

SYMPOSIUM No. 32

ORGANIZED BY THE IAU IN COOPERATION WITH IUGG

HELD IN STRESA, ITALY, 21 TO 25 MARCH 1967

CONTINENTAL DRIFT, SECULAR MOTION OF THE POLE, AND ROTATION OF THE EARTH

EDITED BY

WM. MARKOWITZ

(Marquette University)

AND

B. GUINOT

(Paris Observatory)

Springer-Science+Business Media, B.V.

Published on behalf of
the International Astronomical Union
by
D. Reidel Publishing Company, Dordrecht, Holland

ISBN 978-94-010-3283-4   ISBN 978-94-010-3281-0 (eBook)
DOI 10.1007/978-94-010-3281-0

CONTENTS

Part I

ABBREVIATIONS USED IN THE INTRODUCTION

AGU	American Geophysical Union
BIH	Bureau International de l'Heure
FAGS	Federation of Astronomical and Geophysical Services
IAG	International Association of Geodesy
IAU	International Astronomical Union
ICSU	International Council of Scientific Unions
ILS	International Latitude Service
IPMS	International Polar Motion Service
IUGG	International Union of Geodesy and Geophysics
PZT	Photographic Zenith Tube
UNESCO	United Nations Educational, Scientific and Cultural Organization

INTRODUCTION

The hypothesis of continental drift has become of increasing interest to geophysicists in recent years. The IUGG Upper Mantle Committee has stated that the hypothesis of continental drift envisages horizontal displacements of the continents over thousands of kilometers, and that it is a principal objective of the Upper Mantle Project to prove whether or not continental drift has occurred. The origin of the hypothesis may be traced to the close similarity in outlines of the coasts on the two sides of the Atlantic Ocean. The theory that the eastern and western hemispheres are drifting apart was expounded in particular by A. Wegener.

Modern geophysical theories seek to explain paleomagnetic observations by assuming that two things have occurred in the past: (a) large-scale polar wandering, and (b) continental drift. A direct confirmation of drift, if it exists, is greatly desired.

Attempts have been made to prove that continental drift has occurred from observed changes in latitude and longitude. It was thought by some, in fact, that such changes had been detected. It became apparent, however, that the observed changes could be due to errors of observation. It is now clear that the rate of continental drift is so small, if not zero, that systematic errors must be eliminated, in particular, those due to errors in proper motions of the observed stars. If two stations utilize different star programs then any relative error in the proper motions will give a fictitious relative drift between the stations. Furthermore, changes in latitude and longitude due to continental drift must be separated from changes due to the secular motion of the pole.

The Second Symposium on Recent Crustal Movements, sponsored by the IAG of the IUGG and held at Aulanko, Finland in August, 1965, included several astronomical papers concerned with continental drift. It became evident to Dr. B. Guinot and the writer that new astronomical programs were required to study continental drift effectively. These programs would utilize existing stations for the most part, organized into chains of two or more on nearly the same parallel of latitude, which would observe the same stars. This practice has been employed by the ILS chain on latitude $+39° 8'$ since 1899. No attempt would be made to utilize heterogeneous observations from a large number of stations.

Since there is not sufficient time at the General Assemblies of the IAU and the IUGG to study the scientific problems and to arrange specific cooperative programs it was thought desirable to hold a symposium, to be sponsored jointly by the IAU and IUGG, about five or six months before the holding of the General Assemblies in 1967.

Brigadier Guy Bomford, President of the IAG, who was present at Aulanko, gave his support to this idea. A proposal was submitted to the Executive Committees of the IAU and the IUGG, which gave their warm approval and offers of support.

It was decided that the symposium would be small in character. Its most important function would be to operate as a Working Group, to arrange cooperative observing programs by those concerned. Scientific papers concerned with related astronomical problems, such as the secular motion of the pole and the rotation of the earth, would be included. In addition, some geophysical discussion was planned, in particular, estimates of the expected rate of continental drift. Finally, the symposium would include a discussion on the possible future use of satellite techniques. No attempt would be made to decide whether continental drift was occurring.

In October 1966 Brigadier Bomford suggested that the selection of the minor axis of the earth, which geodesists required, might be considered at the symposium. This, in effect, would be to adopt a fixed origin of reference for the instantaneous coordinates of the pole of rotation. Accordingly, the second major objective of the symposium was to adopt a recommendation on the selection of such an origin.

Opening Session

The Symposium was opened formally on 21 March 1967 by Prof. F. Zagar. The Mayor of Stresa, Dr. Ing. G. Cattaneo, welcomed the participants. Remarks were then made on behalf of the Italian Geodetic Commission by Prof. P. Dore, the President; the IAU by Mr. D. H. Sadler, the Past General Secretary; the IUGG by Dr. G. D. Garland, the General Secretary and the IAG by Prof. P. Tardi, the Past General Secretary. Dr. Wm. Markowitz concluded the opening session by outlining the objectives of the Symposium.

Working Group

This consisted of B. Guinot, Chairman; H. J. Abraham, L. Arbey, J. Bonanomi, H. Enslin, R. G. Hall, D. Monger, A. Orte, A. C. Scheepmaker, H. M. Smith, R. W. Tanner, M. Torao, and S. Yumi.

The Group noted that to obtain observations useful for the study of continental drift it was necessary to eliminate the effect of errors in star positions and proper motions. This can be done by having a common observing program at stations which form a chain. If two stations of a PZT chain are 15' apart in latitude then about $\frac{1}{2}$ the stars can be observed in common, which suffices to link the stations. Astrolabes, however, can be several degrees apart in latitude. Also, the Danjon astrolabe being a portable instrument, it was noted that observations might be made in 10-year cycles, with each instrument of a chain remaining on the same latitude for two years. It was estimated that relative latitudes and longitudes can be determined to about 1 m with a pair of PZT's in one year, and about 2 m with a pair of astrolabes.

One PZT chain is in existence, Mizusawa–Washington, at latitude $+39°$. Calgary–Herstmonceux, at $+51°$, and LaPlata–Mount Stromlo, at $-35°$, are PZT chains scheduled to begin in 1968. These chains could be strengthened or new ones formed by moving existing PZT's. It was noted that an Ottawa–Pulkova chain could be formed at +45° and that the Hamburg and Potsdam PZT's could be moved to the Calgary–Herstmonceux latitude.

Astrolabe chains in operation are: Mizusawa–Algiers, to which Washington can be added, at +36°, and Cape–Santiago, at −34°. Astrolabes also in operation include those at Richmond, Quito, and Sao Paulo. Others to be placed in operation will be at Milan, San Fernando, San Juan (Argentina), and Tierra del Fuego. These can be used to form additional chains.

The Working Group noted that zenith telescopes should be on the same parallel of latitude and that the only such chain existing is that of the IPMS, which determines the fundamental reference system of the polar motion. Meridian-passage instruments are not suitable for the study of continental drift.

The importance of having a central agency to discuss the results obtained from the cooperative programs was stressed.

SELECTION OF ORIGIN OF POLE

The numerical values of the coordinates of the pole, x and y, which are obtained from a group of stations, depend upon the observed variations in latitude, which in turn depend upon the adopted initial latitudes of the stations. The successive Central Bureaus of the ILS have used different initial coordinates, in effect different origins – sometimes several, for different 6-year intervals. This introduced an element of confusion for geodesists, who wished to refer their observations to the same origin.

At the Dublin meeting of the IAU, in 1955, it was recommended that observations for time be corrected for the motion of the pole, in a uniform manner. A Rapid Latitude Service to be conducted by the BIH was established. The polar motion deduced rapidly by the RLS was to be corrected systematically so as to agree with that obtained by the ILS, that is, the same origin would be used. This practice was not strictly followed; data were not obtained rapidly enough in the early stages.

The origin used from 1949 to 1959 by G. Cecchini, the Director of the Central Bureau of the ILS, was called by him the "system of Wanach, 1900–05". In 1959 Cecchini adopted an origin called the "new system, 1900–05" and which is now designated the "mean pole of 1903·0". The fixed, initial latitudes which define this origin are given in Resolution No. 1 adopted at Stresa. This origin was retained when the Central Bureau of the ILS was moved to Mizusawa in January 1962 and the name was changed to Central Bureau of the IPMS.

Meanwhile, the origin used for the computation of $\Delta\lambda$ by the BIH was moved in the interval 1958·75 to 1958·95 to the so-called "mean pole of epoch", which approximates the moving center of the observed polar motion. Thus, different origins were used at different times, and not even the same one by the ILS/IPMS and the BIH. This situation caused difficulties in the reduction of astronomical, geodetic, and satellite observations. Indeed, one could not be certain of the corrections to be used to refer observations to a single origin. Requests therefore arose for the adoption of a single, fixed origin to be used by both the BIH and IPMS.

The discussions at Stresa made clear the desire to adopt a fixed origin. One letter on this subject was received, from Mr. Bruce Lambert of the Division of National Mapping, Australia, who suggested adopting the pole of 1962·0. There were reasons, however, for adopting the pole of 1903·0. It is the one used by the IPMS; also, the Directing Board of the BIH, meeting in Paris on 18 March 1967, had adopted a recommendation that the BIH use the pole of 1903·0. The Stresa Symposium recommended adoption of this origin also.

Resolutions

The resolutions adopted at Stresa were forwarded to the General Secretaries of the IAU and IUGG for consideration at the General Assemblies to be held in the fall of 1967. A brief explanation of the relations between some of the organizations concerned may be helpful here.

The BIH and IPMS are sponsored by the IAU and IUGG and receive financial support from FAGS. FAGS is an agency of ICSU; both receive financial assistance from UNESCO. The IAG is one of the seven Associations which form the IUGG and is the one most closely concerned with the BIH and IPMS. The BIH and IPMS are advised by Scientific Councils whose members include representatives of the IAU and the IUGG. The Directing Board of the BIH is the BIH Scientific Council. The Scientific Councils are guided on general scientific matters by recommendations of the IAU and IUGG. Recommendations adopted by the Scientific Councils, as well as the Symposium at Stresa, will be reported to the IAU and IUGG for their consideration.

Scientific Sessions

There were seven scientific sessions. The subjects and the Chairmen were as follows:
1. Geophysical estimates of rate of drift (P. J. Melchior).
2. Secular motion of the pole and secular changes in latitude and longitude from astronomical observations (H. M. Smith).
3. Continuation of topics of session 2 (M. Torao).
4. Polar motion from observations for time; Rotation of the earth (R. W. Tanner).
5. Artificial satellite techniques (D. H. Sadler).
6. Theoretical papers; Resolutions (J. Bonanomi).
7. Informal reports; Report of Working Group (B. Guinot).

The scientific sessions were opened with an invited paper by G. D. Garland on estimates of continental drift to be expected. In the discussion which followed, Dr. Garland was asked to provide, also, arguments against the theory of continental drift, which he did. The general feeling which resulted from the paper and the discussion was that continental drift of a few centimeters per year might be occurring and that it was worthwhile to attempt to detect this by astronomical observation.

The second and third sessions concerned attempts to disentangle the secular motion of the pole from crustal displacements of observing stations. The discussions concerned chiefly whether past astronomical observations show that crustal displacements had occurred. The arguments for and against are given in the papers.

The fourth session included two papers on the determination of the polar motion from observations for time and two theoretical papers relating to the variable speed of rotation of the earth and the secular motion of the pole.

The possible use of three artificial-satellite techniques for studying continental drift was discussed in the fifth session. The accuracy of a position obtained from optical tracking is now about 30 m. J. A. Weightman thought, based on a study of the Western European Satellite Triangulation Program, that an accuracy of 3 m might be obtained at the end of a 3- to 5-year program. The radio-tracking method described by D. W. Trask and C. J. Vegos gave an accuracy of 10 m for the difference in longitude between two stations and an expected accuracy of 1 m for the future. Latitude differences are not determined by this method. The discussions indicated that the accuracies to be obtained from the optical and radio tracking would not be good enough to compete with the classical astronomical methods.

Interest was therefore centered on the proposal of C. O. Alley and P. L. Bender to place corner reflectors on the Moon and use lasers for ranging. This method would provide a direct distance measurement, instead of an angular separation, between stations. The estimated accuracy is 15 cm. It was agreed that if this method were successful it would provide the most rapid method of detecting continental drift, possibly within 10 years as against 30 to 50 years by the classical methods.

The sixth and seventh sessions included theoretical papers and a few informal reports. T. Nicolini gave the definitive ILS coordinates of the pole for 1941·0 to 1949·0 for the first time. These had been long awaited.

Scientific Papers

A total of 19 papers was given. In accordance with IAU instructions the papers printed here have been shortened where possible, in particular, if a similar, more extensive paper has been or will be published elsewhere. The papers were reviewed by the Editors at Stresa and returned to the authors for revision. In some cases the revised papers printed here take account of the discussions held at Stresa.

Organization of Symposium

The Organizing Committee consisted of Wm. Markowitz (President), E. P. Fedorov, B. Guinot, P. Melchior, M. Torao, and F. Zagar (Chairman, Local Committee).

The Symposium was held in Stresa, Italy, 21 to 25 March 1967, upon invitation of the Italian Geodetic Commission. The sessions were held at the Hotel La Palma, Stresa, which also lodged the participants. This saved time and eliminated a transportation problem.

The program was left uncrowded so as to leave ample time for discussion of the papers and resolutions and for the conferences of the Working Group. There were no formal evening sessions. The instructions contained in the *Astronomer's Handbook* (*Trans. IAU*, Vol. 12C, 1966) were followed in organizing the Symposium and in editing the *Proceedings*.

The IAU and the IUGG jointly sponsored the Symposium and provided the principal financial grants, used chiefly for support of travel of participants. The IAU also undertook the publication of the *Proceedings* of the Symposium. A grant from the American Geophysical Union was used for expenses of organization. The Italian Geodetic Commission contributed for expenses of the Local Committee.

A total of 49 participants from 14 countries attended. The participants expressed regret that their colleagues G. Cecchini, E.P. Fedorov and N. Sekiguchi, who had contributed much to the study of the polar motion, were unable to attend.

Acknowledgments

On behalf of the participants of the Stresa Symposium, I thank the Organizations and the individuals whose interest and support made the Symposium possible. Dr. J.-C. Pecker and Dr. G.D. Garland, the respective General Secretaries of the IAU and IUGG, actively supported the Symposium. Dr. Luboš Perek, the Assistant General Secretary of the IAU, provided guidance for the organization of the Symposium and the preparation of the *Proceedings*. Dr. Waldo Smith, the Executive Director of the AGU, aided in obtaining a grant. The support of Prof. P. Dore was greatly appreciated.

Excursions and reception were kindly provided by the Mayor of Stresa, by the Italian Geodetic Commission, by the Local Committee, and by the Rector of the University of Milan, Prof. G. Polvani.

Everything possible for the smooth running of the meeting and the comfort of the participants at Stresa was done by Prof. F. Zagar and Miss T. Zagar, with the help of the efficient staff from the Brera Observatory.

Dr. E.P. Clancy served as Reporter.

I add, finally, my thanks to the officials of Marquette University for their support in carrying out my own tasks.

Conclusion

The Stresa Symposium had two objectives, which were attained. It is hoped that its work will prove to be in some measure as useful to astronomy and geodesy as was the conference held in 1883, also in Italy, which resulted later in the creation of the International Latitude Service.

WM. MARKOWITZ

President, Organizing Committee

RAPPORT DU GROUPE DE TRAVAIL SUR
LES PROGRAMMES ASSOCIÉS

Les mesures de l'heure et de la latitude ne peuvent donner des résultats incontestables sur la dérive des continents et la dérive du pôle que si les programmes sont associés de façon à éliminer l'influence des erreurs des catalogues d'étoiles.

1. Usage des lunettes photographiques zénithales (PZT)

Les PZT doivent être groupés en chaines sur des parallèles communs. Les instruments d'une chaîne doivent observer les mêmes listes d'étoiles aux mêmes époques; les positions et les mouvements propres des étoiles seront identiques. On s'assurera de plus, par des échanges occasionnels de données d'observations, que les méthodes de réduction sont équivalentes. On veillera à ne pas altérer les sites afin de ne pas modifier la réfraction.

Ces précautions étant prises, on peut estimer qu'une dérive en longitude de l'ordre de 1 m pourrait être révélée par un couple d'instruments, ainsi que des variations des différences de latitude de l'ordre de 1 m.

Les chaînes suivantes sont réalisées ou envisagées:

a. Mizusawa–Washington (réalisée). Seule la moitié des étoiles sont communes. Latitude $+39°$, écart en longitude $9·5$ h.

b. Ottawa (Calgary)–Herstmonceux (débutera en 1968). Latitude $+51°$, écart en longitude 5 h.

c. La Plata–Mount Stromlo (débutera en 1968). Latitude $-35°$, écart en longitude 10 h.

On pourrait étudier la possibilité

d. de joindre les PZT de Hambourg et de Potsdam à la chaîne b.

e. d'établir le PZT de Poulkovo (dont le mouvement est envisagé) à la latitude du nouvel observatoire d'Ottawa; latitude $+45°$.

Si de nouveaux PZT doivent être installés, on recommande qu'ils soient conjugés avec des PZT existants.

2. Usage des astrolabes

A. CHAÎNES D'ASTROLABES SUR PARALLÈLES COMMUNS

Même si les latitudes de deux astrolabes différent de quelques degrés, les programmes peuvent comprendre la plupart des étoiles en commun. Les étoiles non-communes peuvent faire l'objet d'observations hors-programme pour améliorer la liaison. Les méthodes de réduction doivent être mutuellement contrôlées. On peut ainsi espérer mettre en évidence des dérives de l'ordre de 2 m en longitude ou en différence de latitudes.

L'astrolabe étant aisément transportable, on recommande l'usage d'un instrument itinérant qui viendrait se placer successivement sur le parallèle des stations existantes avec un cycle de période de l'ordre de dix ans. La durée des observations en chacune des stations temporaires devrait être de deux ans.

B. RÉSEAU D'ASTROLABES

Comme les astrolabes observent sur une zone de déclinaisons d'environ 55° de large et qu'ils contribuent à améliorer le catalogue d'étoiles fondamental sur toute cette zone, un réseau d'astrolabes à toutes latitudes peut réaliser un catalogue d'étoiles homogène où les erreurs des positions et des mouvements propres seraient très réduites. Ainsi, la variation de l'angle des verticales de points quelconques peut être mesurée avec une erreur correspondant à 2 m sur le sol.

Pour rendre cette méthode efficace, les précautions suivantes doivent être prises:

1. les programmes doivent porter sur les étoiles du FK4.

2. des observations supplémentaires doivent être faites pour améliorer les positions du plus grand nombre possible des étoiles du FK4.

3. les programmes doivent être établis avec le souci d'améliorer au mieux le FK4.

C. RÉALISATIONS ET PROJETS

Il existe des chaînes sur parallèles quasi-communs

a. Mizusawa–Alger (à laquelle pourrait se joindre Washington) Latitude +36° environ, écart en longitude 9 h.

b. Le Cap–Santiago du Chili. Latitude −34°, écart en longitude 6 h.

Le réseau d'astrolabes s'est dévelopé récemment. Outre le groupe d'astrolabes européens, des astrolabes fonctionnent régulièrement à Mizusawa, Alger, Richmond (Floride), Quito, Le Cap, Santiago du Chili, Sao Paulo.

Quelques astrolabes nouveaux doivent être mis en service: à Milan, San Fernando (Espagne), Istanbul, San Juan (Argentine) et à la Terre de Feu (Argentine, latitude −54°).

On note que tous ces astrolabes réalisent une bonne liaison Europe–Amérique du Sud. Il faut cependant encourager de nouvelles observations dans la zone équatoriale pour améliorer la liaison des catalogues d'étoiles des hémisphères boréal et austral.

3. Autres instruments

A. LUNETTES ZÉNITHALES DES LATITUDES

Ces instruments doivent être situés sur des parallèles communs. La seule chaîne réalisée est celle des stations internationales du SIMP qui constituent le système de référence fondamental de la polhodie.

B. INSTRUMENTS DES PASSAGES MÉRIDIENS

Ces instruments sont sujets à des erreurs systématiques trop importantes pour qu'ils puissent être utiles dans l'étude des dérives continentales.

4. Discussion des résultats

On recommande qu'un service soit désigné pour centraliser et discuter les résultats des observations des instruments à programmes associés.

RESOLUTIONS

Recommendations Adopted for Communication to the IAU and IUGG

No. 1. The participants in the above symposium recommend that, for all astronomical and geodetic usage, the coordinates of the pole be referred, as origin of coordinates, to the "mean pole of 1903·0", as defined below.

The "mean pole of 1903·0" is identical with that known as the "new system, 1900–05" of G. Cecchini; it is defined in practice by the following adopted values of the latitudes of the five ILS stations for 1903·0:

Mizusawa	$+39° 08' 03''.602$
Kitab	$01''.850$
Carloforte	$08''.941$
Gaithersburg	$13''.202$
Ukiah	$12''.096$

This recommendation will be communicated to the IAU and the IUGG for their consideration, with the request that each Union should, if it approves the recommendation, take appropriate action to ensure its implementation.

No. 2. (a) Considering that continental drift and the secular motion of the pole can be studied free of the effects of errors in stellar proper motions by suitably organizing the observing programmes of either PZT's or astrolabes located on nearly the same parallel of latitude:

the participants in the above Symposium recommend that:

(I) chains of two or more instruments, either PZT's or astrolabes, be established on nearly the same parallel of latitude, through the cooperation of existing stations wherever possible;

(II) when new stations are established, these be placed in locations which will allow chains to be formed.

(b) The participants in the above Symposium also recommend that a regular distribution of astrolabes in latitude be maintained in order to study the drifts between stations in different latitudes.

No. 3. The participants in the above Symposium recommend that, in the selection of new sites for observing stations, due regard be given to the local stability of the crust and the direction of gravity as evidenced by suitable geophysical measurements; they also draw attention to the fact that similar information for the existing stations would be welcome.

No. 4. The participants in the above Symposium invite attention to the fact that the utilization of the Moon and artificial satellites and of laser techniques may permit the determination of intercontinental distances, and changes therein, with high precision for the study of continental drift.

Note on the Resolutions

The resolutions given above are of two types:

Resolution No. 1 is a scientific resolution which recommends the adoption of specific constants by the IAU and IUGG. These organizations can take such actions as they wish to ensure use of these constants.

Resolutions Nos. 2, 3, and 4 are of an organizational nature, which can be brought to the attention of interested national organizations.

LIST OF PARTICIPANTS

Abraham, H. J., Mount Stromlo Observatory, Canberra, Australia.

Alley, C. O., University of Maryland, College Park, Md. 20740, U.S.A.

Arbey, L., Observatoire Astronomique de Besançon, France.

Bender, P. L., Joint Inst. for Lab. Astrophysics, University of Colorado, Boulder, Colo. 80302, U.S.A.

Billaud, G., Observatoire de Paris, Paris 14, France.

Bonanomi, J., Observatoire de Neuchâtel, Switzerland.

Cahierre, L., IAG, 19 rue Auber, Paris 9, France.

Caprioli, G., Osservatorio Astronomico di Monte Mario, Rome, Italy.

Desio, A., Istituto di Geologia, Università di Milano, Italy.

Dore, P., Facoltà di Ingegneria, Bologna, Italy.

Enslin, H., Deutsches Hydro. Inst., Hamburg, Germany.

Fichera, E., Osservatorio Astronomico di Capodimonte, Naples, Italy.

Fleckenstein, O. J., Osservatorio Astronomico di Brera, Milan, Italy.

Fracastoro, M. G., Osservatorio Astronomico di Pino Torinese, Turin, Italy.

Garland, G. D., University of Toronto, Toronto 5, Canada.

Gougenheim, A., 30 Boulevard Flandrin, Paris 16, France.

Guinot, B., Observatoire de Paris, Paris 14, France.

Hall, R. G., U.S. Naval Observatory, Washington D.C. 20390, U.S.A.

Lippold, H. R., U.S. Coast and Geodetic Survey, Rockville, Md. 20852, U.S.A.

Losert, W., Bund für Vermess., 1082 Schmidtpl. 3, Vienna, Austria.

Markowitz, W., Marquette University, Milwaukee, Wis. 53233, U.S.A.

Marussi, A. E., Università di Trieste, Italy.

Melchior, P., Observatoire Royal de Belgique, Brussels 18, Belgium.

Monger, D., U.S. Naval Observatory, Box 757, Perrine, Fla. 33157, U.S.A.

Morelli, C., Osservatorio Geofisico, Trieste 116, Italy.

Mueller, I., Ohio State University, Columbus, Ohio 43210, U.S.A.

Nicolini, T., Osservatorio Astronomico di Capodimonte, Naples, Italy.

Orte, A., Observatorio de San Fernando, Cádiz, Spain.

Proverbio, E., Osservatorio Astronomico di Brera, Milan, Italy.

Pugliano, A., Osservatorio Astronomico di Capodimonte, Naples, Italy.

Robbins, A.R., Oxford University, 62 Banbury Road, Oxford, England.

Runcorn, S. K., The University, Newcastle upon Tyne, England.

Sadler, D. H., Royal Greenwich Observatory, Herstmonceux, England.

Scheepmaker, A. C., Randweg 201, Bussum, Netherlands.

Schuler, W., Observatoire de Neuchâtel, Switzerland.

Shimizu, T., Kyoto University, Kyoto, Japan.
Smith, H. M., Royal Greenwich Observatory, Herstmonceux, England.
Smriglio, F., Osservatorio Astronomico di Monte Mario, Roma, Italy.
Stoyko, A., Observatoire de Paris, Paris 14, France.
Stoyko, N., Observatoire de Paris, Paris 14, France.
Strand, K. A., U.S. Naval Observatory, Washington, D.C. 20390, U.S.A.
Sugawa, C., International Latitude Observatory, Mizusawa, Japan.
Tanner, R.W., Dominion Observatory, Ottawa, Canada.
Tardi, P., 4 Villa de Ségur, Paris 7, France.
Torao, M., Tokyo Astronomical Observatory, Tokyo, Japan.
Vicente, R., Mestre Aviz 30, R/c, Alges, Portugal.
Weightman, J. A., Geodetic Office, Feltham, Middx., England.
Yumi, S., International Latitude Observatory, Mizusawa, Japan.
Zagar, F., Osservatorio Astronomico di Brera, Milan, Italy.

Staff

Clancy, E.P., Holyoke College, South Hadley, Mass. 01075, U.S.A. *Recorder*
Zagar, Miss T. *Secretary*

Part II

SCIENTIFIC PAPERS

1.1. POSSIBLE RATES OF RELATIVE CONTINENTAL MOTION

G. D. GARLAND

(University of Toronto, Canada)

ABSTRACT

Much of the evidence for continental drift, such as the matching of coastlines, provides only an average rate over a very long time interval. Palaeomagnetic results give average velocities between 0·5 and 3·0 cm/yr, with higher values for some areas such as India. These also represent average rates over long times. Very recent research on the magnetic character of the ocean floors suggest that the present rate of ocean floor spreading can be determined. The continents of South America and Africa may be separating at a rate of 4·0 cm/yr.

RÉSUMÉ

La plupart des témoignages de la dérive des continents donnent seulement une vitesse moyenne correspondant à une période très longue. Les résultats paleomagnétiques donnent des vitesses moyennes entre 0·5 et 3·0 cm/an, avec des valeurs plus grandes pour l'Inde. Les recherches nouvelles sur le caractère magnétique du fond des mers peuvent indiquer la vitesse présente; il est possible que les continents de l'Amérique du Sud et d'Afrique se séparent à 4·0 cm/an.

A great deal has been written during the past five years in support of the hypothesis of continental drift, although objections to it remain. The purpose of the present paper is not to engage in arguments for or against, but simply to review the estimates that are available on the possible rate of continental movement, assuming that it has indeed occurred. These estimates can be deduced from certain of the arguments that have been advanced by authors who have written in favour of the theory.

From the point of view of astronomical tests, however, it is important to distinguish between estimates of average rate over long geological periods, and estimates of present-day velocities. Obviously, it is the latter which are important to any programme of measurements to detect continental displacements within a reasonable number of years, but most of the older estimates give only the former quantity.

The estimates of average rate over a long geological time can be deduced from many of the standard arguments in favour of continental drift. These include:

a. the matching of coastlines, originally by eye and more recently by computer (Bullard *et al.*, 1965);

b. the dispersal of areas of similar climatic history, such as those of Carboniferous glaciation;

c. the displacement of continents as indicated by palaeomagnetic measurements.

All observations of these types give only a displacement, and to determine a rate of movement, a time must be associated with the movement. For example, a. and b. suggest the displacement of South America relative to Africa, a distance of 5500 km, in the time since the close of the Carboniferous (280×10^6 years). A minimum rate of movement is therefore 2·0 cm per year, but this evidence by itself does not preclude the possibility that the displacement was accomplished in much more recent time with a correspondingly greater drift rate. In fact, it has been suggested on geological grounds that most of the separation of Africa and America took place in the last 150×10^6 years. Since the rebirth of interest in continental drift owes much to the results of palaeomagnetic measurements, it is desirable to renew the nature of the observations and their role in establishing a rate in some detail.

Measurements of the vector-magnetic intensity of either a sedimentary or an igneous rock are assumed to give the direction of the earth's magnetic field at the time of formation of the rock. Much has been written on the sources of possible complicating factors (see, for example, Runcorn, 1962) but it is now generally accepted that anomalous cases (unstable magnetization, subsequent chemical alteration, etc.) can be recognized, and that reliable determinations of the direction of the field are available at many points for a series of times over at least the past 500 million years. If the dipole nature of the earth's field is assumed, the inclination of the magnetization vector in the sample gives, uniquely, the original magnetic latitude of the site. The original longitude of the site is indeterminate, and the azimuth toward the ancient pole cannot be determined if the rock mass from which the sample was obtained suffered any rotation during its history. If it is assumed that no rotation has occurred, the palaeomagnetic measurements yield an ancient pole position. Measurements from samples of different geological age give a "polar-wandering curve", leading toward the present north magnetic pole for very young samples (Figure 1). A single polar-wandering curve indicates only the possible motion of the poles relative to the earth's surface as a whole. Evidence for continental drift is obtained only when discrete curves are obtained from samples collected from different curves, as shown also in Figure 1. The fact that these curves are displaced from each other is evidence for continental displacement since the date of formation of the rocks on which curves are based. The curves thus converge toward the present pole, and the rate of convergence, in a sense, indicates a rate of drift. The westerly displacement of the Americas from Europe and Africa is indicated in Figure 1, but even more striking are the curves for Australia and Antarctica, which suggest very large displacements for these continents. Since each point on any of the polar-wandering curves is simply the centre of a rather large region of uncertainty, it is difficult at present to use the displacement of the curves to determine a meaningful drift rate. Estimates of drift rate have been made (Deutsch, 1966) from the change in latitude of samples alone, since, as noted above, this is the quantity best determined by palaeomagnetic measurements. Once again, the rate obtained is simply an average over the whole time since the rock was formed, and there is no way of telling if the change in latitude was accomplished in a much shorter time. Considering samples from six land masses, excluding India, Deutsch obtains a range of average drift rates, in latitude only, of 0·5–2·5 cm/year, in the time since Carboniferous. When very young rocks (less than 25×10^6 years old) alone are considered, the indicated rates of drift in latitude range from 0 to 3 cm/year. The case of India is of particular interest, as there is strong evidence for a northward drift of that landmass in the past 60 million years, indicating a rate of 4–12 cm/year.

Most of the estimates of drift rate over shorter periods, which are probably more appropriate to the present day, are related to observations of the ocean floor. There is

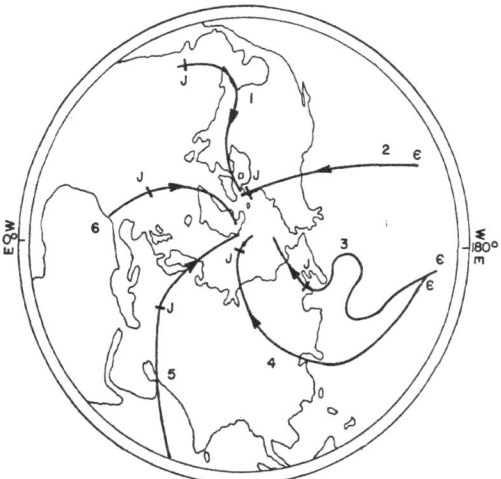

FIG. 1.   *Examples of polar-wandering curves (after Deutsch). Curves are numbered according to location of samples; 1. India; 2. Africa; 3. W. Europe; 4. N. America; 5. Antarctica; 6. Australia. The letters $\in$ and J correspond to Cambrian and Jurassic times respectively.*

considerable evidence (Dietz, 1961) that the floors of oceans are created from basaltic material injected along the oceanic ridges (Figure 2), and then carried to either side. Although it is not the intention here to discuss mechanisms, it may help to visualize the process if one thinks of the oceanic crust subjected to forces provided by mantle-convection cells, with currents rising under the ridges, and dragging the oceanic crust horizontally toward the loci of downward currents. Wilson (1963) has noted that the isotopic ages of oceanic islands increase with distance from the ridges and has proposed that they were all formed at the ridge, and swept aside. He has produced a graph of age versus distance which indicates a horizontal velocity of 3.5 cm/year. There is considerable scatter to the points on the graph, but the rate does appear to be appropriate to within a few million years of the present time.

Wilson (1966) and others have also suggested that rift valleys, such as the Red Sea or the East African rifts represent an earlier stage in the process of oceanic spreading.

If this is the case, The Red Sea represents an opening of 500 km in 25×10^6 years, giving a rate of separation of 2 cm/year. The corresponding figures for the East African Rifts are 50 km in 12×10^6 years or a separation of 0·4 cm/year.

However, it now appears that there is a much more precise method to study the spreading of the ocean floor, and that is by means of the magnetic character of the oceanic crust. Air-borne and ship-borne magnetometer surveys of the oceans have revealed a remarkably linear pattern of anomalies in the magnetic field, with strips of alternately high and low field. The amplitude of these anomalies is such

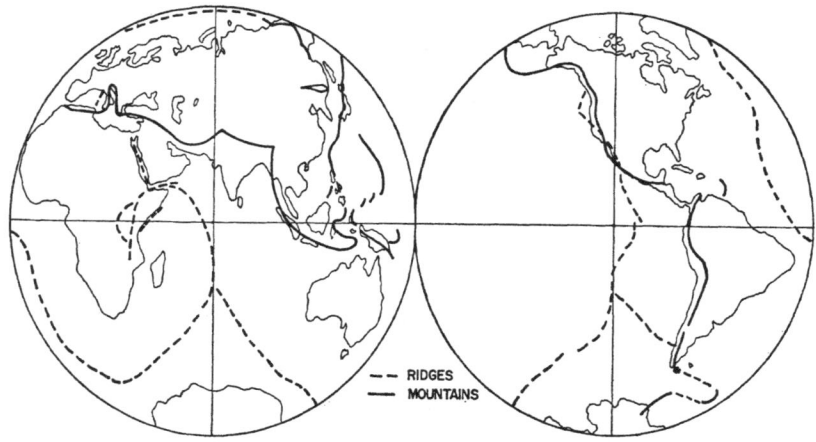

--- RIDGES
— MOUNTAINS

FIG. 2. *The world ridge system.*

that it would be very difficult to explain them in terms of differences in the suscep-tibility of the oceanic crust. They appear to require the presence of bands of rock, with permanent magnetization alternating between the normal and reversed direc-tion. The fact that the bands are roughly parallel to oceanic ridges led Vine and Matthews (1963) to suggest that they represent rock injected at the ridges, polarized in the direction of the magnetic field at the time of emplacement, and subsequently moved aside. Strong support to this hypothesis is lent by the remarkable symmetry of the pattern on the two sides of a ridge (Figure 3) and by the fact that the widths of bands correlate with the varying lengths of periods of reversals in the earth's main field. The time scale of reversals for the past four million years has been established independently from studies on other rocks, and this indicates four major periods with at least three short-period "events" superimposed on them. The pattern in the vicinity of the ridges is in striking agreement with this timescale.

The power of this method is that it provides a spreading rate up to very recent times – virtually the present rate of spreading, as contrasted with other methods, and that it can be applied to different parts of the oceans in turn. Pitman and Heirtzler (1966) have shown that the spreading rate on either side of the Pacific–Antarctic ridge is 4·5 cm/year, in contrast to a rate of 1 cm/year for the Atlantic Ocean south of Iceland. There is evidence that the spreading rate for the South Atlantic is perhaps twice as great as for the North, as though the Atlantic were opening with a pivotal motion, the pivot being north of Iceland.

There is reason to believe, therefore, that a most powerful method is available for determining the present rate of ocean-floor spreading. However, for our present purpose,

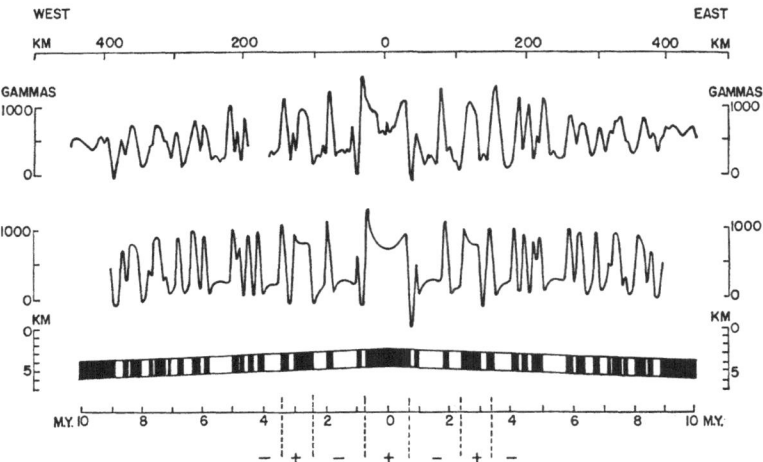

FIG. 3. *Magnetic profile observed (upper) across the Pacific Antarctic Ridge. The lower curve is computed for the assumed ocean-floor structure of normally – and reversely – magnetized blocks shown below it. The time scale for the four most recent epochs of geomagnetic polarity is also shown. (After Pitman and Heirtzler.)*

a careful distinction must be made between ocean-floor spreading and relative motions of the continents. That the two motions are different is obvious when one considers the case of South America: evidently, both the South Atlantic and South-Eastern Pacific Oceans are spreading, but the continent cannot be moving west and east at the same time. The explanation is that the spreading oceanic crust must, in some places, be dragged under the continents, so that continental masses ride over it. It appears that this down-dragging takes place along the line of oceanic trenches, these being assumed to represent the loci of downward convection currents. If this is the case, continents on either side of an ocean with a spreading floor may not move apart if there is such a feature intervening. There is a well-defined trench along the west coast of South America, but there is not in the Atlantic between South America and Africa. It may therefore be assumed that Africa and South America are separating at a rate equal to twice the spreading velocity appropriate to the South Atlantic. This would give a rate of separation of as much as 4 cm/year, assuming that the spreading velocity is twice that found south of Iceland.

Magnetic surveys of the ocean floors are being extended at present by a number of groups, and may be expected to suggest other areas of rapid spreading. On the basis of presently-available evidence the pattern of movement in the Indian Ocean appears to be complicated. For the direct testing of continental drift, therefore, one probably cannot do better at the moment than to take sites on either side of the South Atlantic. Because of the pivotal nature of the motion, mentioned above, a gain of a factor of 2 or 3 is achieved over the rate of separation in the North Atlantic, and this could well make the difference between a detectable or non-detectable displacement in a reasonable number of years.

Measurements in other locations would be useful to test other parts of the hypo-thesis. For example, precise determination of the position of oceanic islands could be used to test the rate of ocean-floor spreading, and islands on either side of the Pacific-Antarctic ridge could be expected to separate at the very high rate of 9 cm/year. Finally, in some areas, such as across the Red Sea, direct geodetic, as opposed to astronomical, measurements could be used to detect separation.

Measurements of this type have already been made by Danish and Canadian groups to test for relative motion between Greenland and the Canadian Arctic islands.

References

Bullard, E.C., Everett, J.E., Smith, A.Gilbert (1965)   *Phil. Trans. R. Soc. Lond.*, **A258**, 41–51.

Deutsch, Ernst R. (1966)   The Rock Magnetic Evidence for Continental Drift, in *Continental Drift*, Ed. G.D. Garland, R. Soc. Canada, Spec. Publ. no. 9.

Dietz, R.S. (1961)   *Nature*, **190**, 859.

Pitman, W.C., Heirtzler, J.R. (1966)   *Science*, **154**, 1164–1171.

Runcorn, S.K. (1962)   Palaeomagnetic Evidence for Continental Drift and its Geophysical Cause, in *Continental Drift*, Ed. S.K. Runcorn, Academic Press, London and New York.

Vine, F.J., Matthews, D.H. (1963)   *Nature*, **199**, 947–949.

Wilson, J.T. (1963),   *Nature*, **197**, 536–538.

Wilson, J.T. (1966)   Some Rules for Continental Drift, in *Continental Drift*, Ed. G.D. Garland, R. Soc. Canada, Spec. Publ. no. 9.

2.1. CONCURRENT ASTRONOMICAL OBSERVATIONS FOR STUDYING CONTINENTAL DRIFT, POLAR MOTION, AND THE ROTATION OF THE EARTH

WM. MARKOWITZ

(Marquette University, Milwaukee, U.S.A.)

ABSTRACT

The analysis of 66 years of concurrent latitude observations of the ILS shows that the mean pole has a secular motion which consists of a progressive component of about $0\overset{''}{.}0035$/yr (10 cm/yr) along the meridian 65°W and a librational component (oscillation) of 24-year period along the meridian 122°W (or 58°E). Crustal displacements in latitude are not found within the errors of observation, about 1 cm/yr.

Comparable, concurrent observations for time (longitude) have not been made but programs are being organized. From 30 to 50 years will be needed for detection of continental drift with PZT's and astrolabes if relative drifts in longitude of 3 cm/yr are occurring.

RÉSUMÉ

L'analyse de 66 années d'observations de latitude par le SIL montre que le pôle moyen a un mouvement séculaire qui consiste en une composante linéaire d'environ $0\overset{''}{.}0035$ par an (10 cm/an) le long du méridien $+65°$ et en une libration de 24 ans de période le long du méridien $+122°$ (ou $-58°$). On n'a pas trouvé de déplacements de la croûte dans la limite des erreurs d'observation, soit environ 1 cm/an. Des observations associées comparables, pour le temps (longitude), n'ont jamais été faites, mais on est en train d'organiser des programmes. 30 à 50 ans seront nécessaires pour mettre en évidence des dérives continentales avec des lunettes photographiques zénithales et des astrolabes, si des dérives en longitude de 3 cm/an se produisent.

1. Introduction

Concurrent observations are defined to be those made by the stations of a chain which observe stars in common. Such observations are important because the polar motion derived from a chain with sufficient stations is independent of errors in the star positions. Three stations suffice if latitude only is determined.

This paper analyzes the concurrent observations made during 66 years, 1900·0 to 1966·0, by the International Latitude Service chain (ILS) at latitude $+39°$. The progressive and 24-year librational components of the secular motion found in 1960 (Markowitz, 1960, 1961) have continued since then. Relative crustal displacements are not found. However, continental drift may be occurring in longitude. No long series of concurrent time (or longitude) observations exists, however.

Concurrent observations in longitude may be obtained with PZT's and astrolabes by revising the observing programs of instruments which are suitably located in latitude. The errors of observation and the time required for detection are discussed.

2. Independence from Declinations Used

It has been stated often that errors in star positions (and proper motions) effect the determination of the pole by the ILS. Let us analyze the equations used to determine the coordinates of the pole, x and y, from three or more stations.

Let δ be the average apparent declination of a star pair and let ζ_i be the average north Zenith distance as observed at station i. Then the observed latitude is

$$\varphi_i = \delta - \zeta_i. \tag{1}$$

Let φ_{i0} be a set of adopted constants, called the initial latitudes. Then, by definition, the variation in latitude is

$$\Delta\varphi_i = \varphi_i - \varphi_{i0}. \tag{2}$$

The equation of condition used in the least squares solution is

$$\Delta\varphi_i = x \cos \lambda_i + y \sin \lambda_i + z, \tag{3}$$

where λ_i is the longitude of station i, and z is a constant. The solutions are

$$x = \Sigma a_i \, \Delta\varphi_i, \quad y = \Sigma b_i \, \Delta\varphi_i, \quad z = \Sigma c_i \, \Delta\varphi_i. \tag{4}$$

The coefficients have the properties, for $i \geq 3$,

$$\Sigma a_i = 0, \quad \Sigma b_i = 0, \quad \Sigma c_i = 1. \tag{5}$$

It can be shown from Equations (2), (4), and (5) that

$$x = A - \Sigma a_i \zeta_i, \quad y = B - \Sigma b_i \zeta_i, \quad z = C - \Sigma c_i \zeta_i, \tag{6}$$

where

$$A = -\Sigma a_i \varphi_{i0}, \quad B = -\Sigma b_i \varphi_{i0}, \quad C = \delta - \Sigma c_i \varphi_{i0}. \tag{7}$$

Hence, x and y do not depend upon the δ's used.

3. Secular Motion of the Pole

The secular motion of the pole previously found consists of a progressive motion of $0\,\rlap{.}''0032$/yr (10 cm/yr) along the meridian 60° West and an oscillating (librational) motion of 24-year period along the meridian 122°W (or 58°E). See page 350 of (1).

The position of the mean pole at 6-year intervals up to 1957·0 as shown in Figure 10 of (1) was based on results published by G. Cecchini. Figure 1 of this paper shows an additional point, for 1963·0, based on results published by Yumi and Wakō (3). The

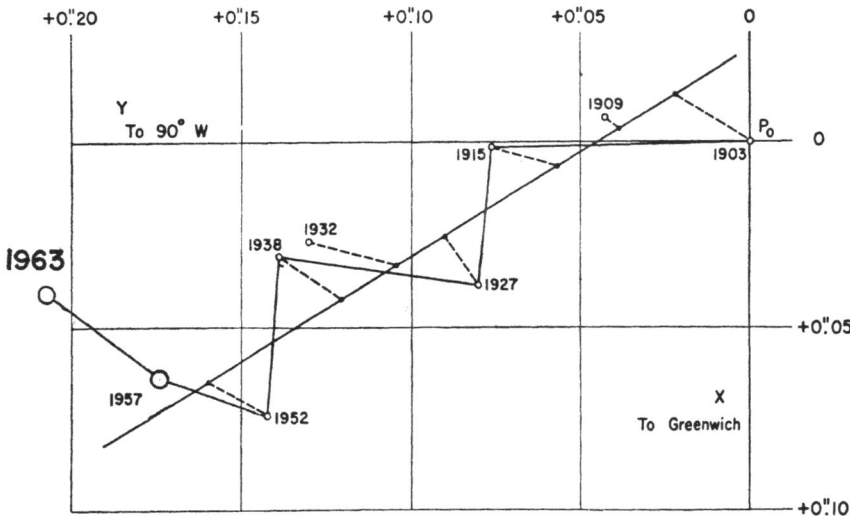

FIG. 1. *Motion of mean pole for 6-year intervals. Origin is pole of 1903·0. The 1963 point was added to a figure drawn previously. The 1957 point was moved slightly.*

location of the 1963·0 point agrees very well with that to be expected from the motions previously found. The presence of the 24-year term is surprising. It is purely empirical, since we do not know of any geophysical phenomenon with this period.

4. Variations in Latitude

Table 1 gives mean values of the $\Delta\varphi_i$, x, y, and z for 6-year intervals for Mizusawa, Kitab, Carloforte, Gaithersburg, and Ukiah. (Kitab was not in operation until about 1930, so that $\Delta\varphi_k$ for 1903 is not an observed value.)

Table 1

Variations in latitude, $\Delta\varphi_i$, and solutions
(Unit $= 0''.001$)

Nominal epoch	M	K	C	G	U	x	y	z
1903·0	0	(0)	0	0	0	0	0	0
27	− 205	− 181	− 85	− 46	− 65	+ 39	+ 80	− 117
32	− 225	− 231	− 104	+ 2	− 17	+ 27	+ 130	− 116
38	− 153	− 123	− 36	+ 116	+ 79	+ 31	+ 139	− 37
52	− 213	− 160	− 13	+ 92	+ 18	+ 74	+ 142	− 64
57	− 277	− 248	− 80	+ 65	− 3	+ 63	+ 173	− 116
63	− 340	− 274	− 131	+ 45	+ 74	+ 40	+ 207	− 132

WM. MARKOWITZ

Table 2

Corrected Variations, $\Delta\varphi_i'$
(Unit = 0″001)

Nominal epoch 1903.0	M 0	K (0)	C 0	G 0	U 0
27	− 88	− 64	+ 32	+ 71	+ 52
32	− 109	− 115	+ 12	+ 118	+ 99
38	− 116	− 86	+ 1	+ 153	+ 116
52	− 149	− 96	+ 51	+ 156	+ 82
57	− 161	− 132	+ 36	+ 181	+ 113
63	− 208	− 142	+ 1	+ 177	+ 206

Essentially, z represents the error in the variation of latitude due to the average error in the star places for the interval. We can correct this by subtracting z. Table 2 gives $\Delta\varphi_i' = \Delta\varphi_i - z$.

The corrected variations in latitude are plotted in Figure 2. The dashed lines are the

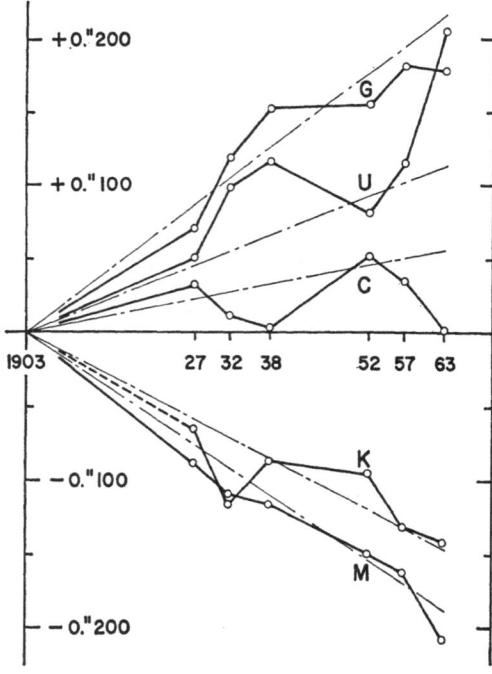

FIG. 2. *Variation in latitude for 6-year intervals. Origin is pole of 1903·0. The ordinates are $\Delta\varphi_i' = \Delta\varphi_i - z$. The dashed lines represent the variations which would have occurred if the mean pole had moved 0″210 in 60 years along longitude 65° W.*

variations which would have occurred if the mean pole had moved $0''.210$ in 60 years along a straight line inclined $+65°$ to the x-axis. The average speed is $0''.0035/yr$ (10 cm/yr).

If the pole moves an amount H in the direction $122°W$ then the latitudes will change by H times the following factors: $M, -\cdot12$; $K, -\cdot99$; $C, -\cdot64$; $G, +\cdot71$; $U, +1\cdot00$. Hence, the librational motion should produce oscillations in the variation of latitude of the same phase for Mizusawa, Kitab and Carloforte, and of the opposite phase for Gaithersburg and Ukiah. Furthermore, the oscillation for Mizusawa should be very small. This is what Figure 2 shows. The small amplitude for Mizusawa is noteworthy, because it has been frequently asserted that this station undergoes crustal displacements.

5. Non-polar Variations

Observed variations in latitude could be due to effects other than polar motion. We divide these into two classes: Type 1, which would necessarily affect all instruments on a station alike, and Type 2, which would not. Examples are:

1a. Crustal displacements.

1b. Changes in direction of gravity.

2a. Changes in refraction.

2b. Changes in properties of instruments.

2c. Changes in observers, or in their personal equations.

Instruments on the same station, such as meridian circles, Zenith telescopes, PZT's, and astrolabes exhibit systematic variations which persist for weeks, months, and longer. Hence, Type 2 effects exist. Type 1 are only conjectured – as yet.

6. Errors of Observation

When comparing observed motions with probable errors to judge whether the pole has moved secularly or whether crustal displacements have occurred it is necessary that external probable errors be used, and not theoretical, internal probable errors. If e_1 is the p.e. for one observation then the theoretical p.e. for n observations, $e_n = e_1/n^{\frac{1}{2}}$, is the correct probable error to use only if the observations are independent. Experience shows, however, that systematic effects generally occur, such as type 2, and the external p.e. is greater than e_n.

I have obtained probable errors for mean $\Delta\varphi_i$ as follows:

Interval	Internal	External
1 night	—	$0''.05$
1 year	$0''.004$	$0''.022$
6 years	$0''.009$	$0''.015$

The external value for one night was obtained from the night to night differences within a month (same groups observed during the month). The external value for one year was obtained from residuals $v_i = (o-c)_i$, published by Yumi and Wako (3). A residual is the difference between the observed $\Delta\varphi_i$ and the computed value obtained by the use of equation (3) and the yearly solutions for x, y, and z. The external value for six years was obtained similarly from Table 1, above. The internal errors were based on the number of nights or the years in the interval.

The p.e. of a coordinate is obtained from equation (4). Using $0\overset{''}{.}015$ for the p.e. of a 6-year mean $\Delta\varphi_i$ we obtain p.e.'s of $0\overset{''}{.}011$ for x and $0\overset{''}{.}009$ for y for a 6-year interval. These values indicate that the progressive and librational motions shown in Figure 1 are not due to accidental causes.

Crustal displacements would be indicated by systematic departures in Figure 2 of the observed variations in latitude from the dashed lines. The existence of the 24-year oscillation of the pole increases the uncertainty of detecting such displacements. A study of the results indicates that crustal displacements in latitude, if they exist, were less than $0\overset{''}{.}02$ in 60 years. This rate is $0\overset{''}{.}00033$/yr, or 1 cm/yr.

7. Choice of Origin

A fixed origin for describing the polar motion is a necessity from a computational standpoint. The ILS pole, which is independent of errors in star positions, provides the fundamental polar motion. Determinations made with other instruments must be adjusted systematically so as to be in accord with the ILS system.

An examination of the equations in Section 2 shows that the origin for the ILS polar motion is defined by the constants φ_{i0}. In 1959 G. Cecchini adopted the following φ_{i0}:

	φ_{0i}	λ_i
Mizusawa	$39° 8'\ 3\overset{''}{.}602$	$-141°$
Kitab	$1\overset{''}{.}850$	$-67°$
Carloforte	$8\overset{''}{.}941$	$-8°$
Gaithersburg	$13\overset{''}{.}202$	$+77°$
Ukiah	$12\overset{''}{.}096$	$+123°$

These define an origin which he called the "new system, 1900–05". It is called the pole of 1903·0 for short. This origin has been widely used for some time, and is a logical choice for the fixed origin.

8. Detection of Continental Drift

The p.e. for one year of observation in latitude by an ILS station is $0\overset{''}{.}022$. I have previously found a p.e. of $0\overset{s}{.}002$ (in time) for one year of observation with a PZT (4).

This corresponds to about 0″.02 along a great circle, so that the probable errors are similar. The equivalent distance is 0·6 m. The probable error of the difference in latitude or longitude for a pair of stations is thus about 1 m.

Although we have not detected continental drift in latitude from the ILS results, such drift may occur in longitude. Assuming a drift rate of 3 cm/yr, it will require 33 years for a drift of 1 m, which corresponds to 0″.03. This is a small quantity to detect definitely, because of possible systematic effects. It might not be possible to be assured that drift had occurred until the total drift was about 2 m or 0″.06. Thus, a chain of PZT's or astrolabes would require about 50 years to detect continental drift if the rate is 3 cm/yr. The time would be diminished, perhaps to 30 years, if a number of chains were in operation. It is not necessary that observations be made continuously. The astrolabe is a portable instrument and it could be used at several latitudes in a cyclic manner.

A complicating factor in detecting continental drift is that secular changes in latitude and longitude due to drift must be separated from those due to the secular motion of the pole. For this reason, continued observations by the ILS are necessary. The preceding consideration does not apply in the satellite methods in which the distances between stations are determined directly, as in the corner-reflector laser method.

9. Speed of Rotation

A comparison of UT2 obtained with the Washington and Richmond PZT's versus atomic time, A.1, shows that the earth's rotation is subject to sudden changes in acceleration but not in speed. Sudden changes in acceleration occurred about September 1957 and January 1962. No correlation with the motion of the mean pole is evident. However, another 20 years of additional observation are needed for more detailed study.

10. Discussion

Numerous attempts have been made to explain the observed secular changes in latitude of the ILS stations by crustal displacements, in whole or in part, rather than by a secular motion of the pole. The stations most frequently supposed to be drifting are Mizusawa and Ukiah. A test of this hypothesis is to compute the polar motion for various combinations of two stations at a time using the data in Table 2. In all cases, including the Carloforte–Gaithersburg combination, which makes no use of either Mizusawa or Ukiah, the motion obtained is similar to that shown in Figure 1. Hence, crustal displacements cannot cause the motion shown by Figure 1.

Yumi and Wako (3) have assigned about half the secular change in latitude of Mizusawa and Ukiah to progressive crustal drifts of opposite sign. They base this hypothesis on the fact that the residuals v_i for the years near 1963 are larger than for the years near 1936. However, the v_i during the 10-year interval from 1950·0 to 1960·0 are the smallest of any similar interval from 1900·0 to 1966·0. If the hypothesis were correct it would mean that from 1950 to 1960 the two stations were back in their initial positions.

Hypotheses of crustal displacement, moreover, cannot account for the agreement (or opposition) of phase shown by Figure 2 and for the small amplitude shown by Mizusawa.

The secular motion of the pole is judged to be real and has been detected because it is relatively large, about 10 cm/yr. Crustal drifts have not been detected because the drift rate in latitude for the ILS stations is small, 1 cm/year or less.

Attempts have been made to prove that secular changes in longitude, not due to polar motion, have occurred. The observations used were non-concurrent. Also, different instruments, star lists, and methods of receiving time signals have been changed from time to time, and instruments were changed in location, as at Washington, Richmond, and Ottawa (5).

Such heterogeneous data cannot be used to detect continental drift. However, concurrent observations for time and latitude, made in the future with PZT's and astrolabes, and combined with continuing ILS observations may detect continental drift in the next 30 to 50 years. There is the possibility of course, that the laser method may detect such drift much sooner.

Additional details of the topics discussed here will be published in the *Bulletin Géodésique*.

References

1. Markowitz, W. (1960)   in *Methods and Techniques in Geophysics*, Ed. S.K. Runcorn, Interscience Publ., New York, p. 325.
2. Markowitz, W. (1961)   *Bull. géodés.*, 59, 29.
3. Yumi, S., Wakō, Y. (1966)   *Publ. int. Latit. Obs. Mizusawa*, 5, 61.
4. Markowitz, W. (1964)   *Proc. Int. Conf. on Chronometry, Lausanne, 1964*, 1, 157.
5. Markowitz, W. (1966)   *Proc. Second Symp. on Recent Crustal Movements, Aug. 1965, Helsinki*, p. 241.

2.2. ON THE SECULAR MOTION OF THE MEAN POLE

SHIGERU YUMI and YASUJIRO WAKŌ

(International Latitude Observatory of Mizusawa, Japan)

ABSTRACT

Analyzing the notable increases in the residual latitudes of the five ILS stations, local drifts of $-0''.00156$/year for Mizusawa and of $+0''.00105$/year for Ukiah were derived. Subtracting the apparent motion of the mean pole due to the local drifts, the secular motion of the mean pole is $0''.00220$/year in the direction $77°.7$ W and this seems to be the real one.

RÉSUMÉ

En analysant les augmentations notables dans les latitudes résiduelles des cinq stations du SIL, on a déduit les dérives locales de $-0''.00156$/an pour Mizusawa et $+0''.00105$/an pour Ukiah. Si on soustrait le mouvement apparent du pôle moyen qui est causé par les dérives locales, le mouvement séculaire du pôle moyen est $0''.00220$/an dans la direction de $77°.7$ W et il semble être le mouvement réel.

Change of the position of the mean pole has hitherto been studied and discussed by many researchers, and the reality of secular change and libration were suggested by Sekiguchi (1), Hattori (2) and Markowitz (3) assuming that the relative positions of the ILS stations did not change with each other. However, there is a supposition that the real change in mean latitude of any station which is expected from a supposed displacement related with a crustal movement, or a change in the plumb line at the station, etc., would produce an apparent change of the mean pole.

The annual mean value of the residual latitudes at each station $(o-c)$, the difference between the observed latitude and the expected one computed from the derived polar coordinates, has a remarkable secular increase except for Carloforte in addition to a large fluctuation from year to year, as shown in Figures 1 and 2. See (4).

It amounts to $0''.03-0''.07$ for the latest years, and the probable error of the derived coordinates of the pole is about $0''.03$, which far exceeds the value of a several thousandths of a second of arc expected from the precision of the monthly mean latitudes and the scattering of the monthly values of $(o-c)$. This suggests the reality of local errors or local secular drift accompanied by fluctuations at any station and the resulting apparent motion of the mean pole. In this paper, however, only the secular part is dealt with, because the available data and number of stations are considered to be too few to discuss the effect of fluctuation in location.

Analyzing the residual latitudes $(o-c)$ of the five ILS stations and comparing them with each other, the secular change in the local error at Mizusawa $(-0''.00312)$ or at

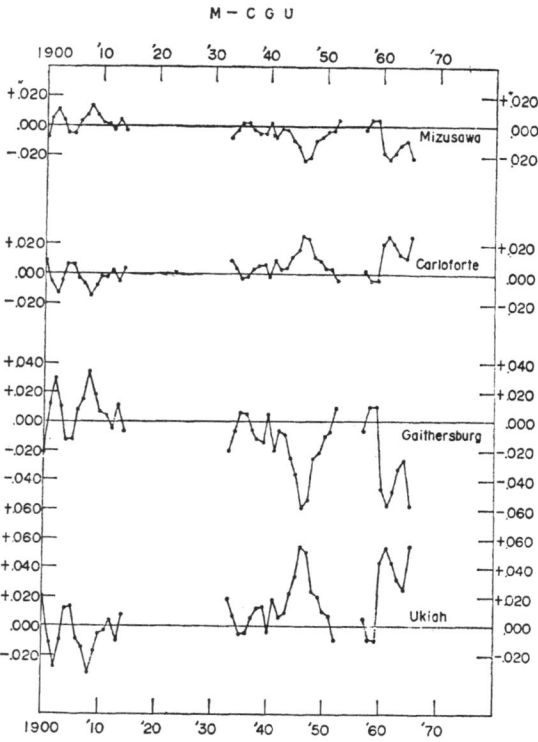

FIG. 1.   *Residuals, o–c, for 4 stations.*

Ukiah ($+0''.00209$) was found to be the main cause of the obtained secular change of (o–c) at each station. Assuming the half and half secular changes, $-0''.00156$/year for Mizusawa and $+0''.00105$/year for Ukiah, the secular changes in the residual latitudes of all the ILS stations may almost vanish. This indicates the possibility of local drift at any station, the amount of which is supposed to be about $0''.001$/year.

This assumption leads to the conclusion that the secular change of the mean pole which has hitherto been obtained from the results of Mizusawa, Kitab, Carloforte, Gaithersburg and Ukiah is reduced to $\frac{1}{2}$ in its x-component and $\frac{3}{4}$ in the y-component. The resulting magnitude of the reduced secular change is

$$\left.\begin{array}{l} \varDelta S_x = +\,0''.00047 \\ \varDelta S_y = +\,0''.00215 \end{array}\right\} \quad \varDelta S = 0''.00220, \quad \theta = 77°.7\ \text{W},$$

and this change seems to be real.

This assumption also leads to another conclusion, that the probable errors of the coordinates of the barycentres for the latest years which amount to about $\pm 0''.03$ in

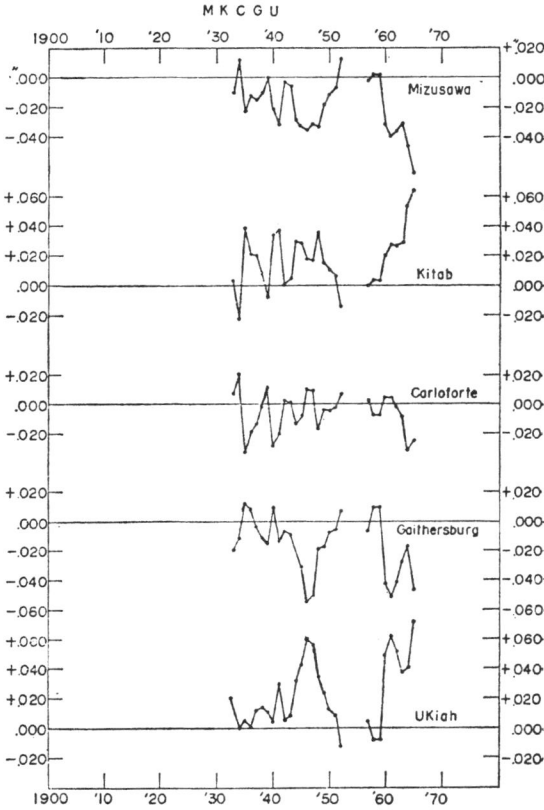

Fig. 2.   *Residuals for 5 stations.*

the current system of MKCGU are reduced to a relatively small amount of $\pm 0''01$, even though this is not quite the satisfactory one.

The treatment of the residual latitudes of the stations also suggests the necessity of some corrections to the mean latitudes in the new system 1900–05, that is $+0''036$ for Mizusawa and $-0''026$ for Ukiah.

In order to detect more precisely the reality of the local drifts of the stations and of the resulting secular motion of the mean pole, additional stations on the same parallel of $+39°8'$ should be established to provide a nearly equal distribution in longitude. For the time being, Tientsin in China and Cincinnati in the U.S.A. are the stations desired. On the other hand, as the astronomical time and latitude are observed referring to the plumb line at the station on the earth crust, change of plumb line and the deformation of the earth crust which might be related with the continental drift should be observed geographically and geodetically in parallel with the astronomical observations.

As far as the latitude is observed independently at any station after its own programme, the mean latitude of the independent station cannot be obtained precisely from the observed latitudes, even corrected by the chain method, but the error in the proper motion will apparently be introduced into the result as the secular variation of the local drift. The errors in declination and proper motion should be determined by meridian circle observations.

Detailed discussions and some comments are given in a recent paper by the present authors in (**4**).

References

1. Sekiguchi, N. (1954)   *Publ. astr. Soc. Japan*, **5**, 109.   Sekiguchi, N. (1956)   *Publ. astr. Soc. Japan*, **8**, 13.
2. Hattori, T. (1959)   *Publ. int. Latit. Obs. Mizusawa*, **III**, No. 1, 1.
3. Markowitz, Wm. (1960)   *Methods and Techniques in Geophysics*, p. 350.
4. Yumi, S., Wakō, Y. (1966)   *Publ. int. Latit. Obs. Mizusawa*, **V**, No. 2, 75–77.

2.3. THE DIRECTION OF THE MINOR AXIS OF GEODETIC REFERENCE SPHEROIDS

BRIGADIER G. BOMFORD and DR. A. R. ROBBINS

(Oxford University, Oxford, England)

ABSTRACT

It is necessary to standardise the definition of the direction of the minor axes of spheroids used for geodetic computations. In view of polar motion, the definition "parallel to the axis of rotation" is ambiguous. Few countries have explicitly stated the direction adopted, and it has generally been implicitly defined by the corrections for polar motion applied to astronomical azimuths, latitudes, and longitudes. But these corrections have often been neglected or applied inconsistently, and errors of perhaps $0''.5$ of azimuth result. The choice for a recommended Geodetic Mean Pole, to which spheroidal minor axes can in future be defined to be parallel, seems to lie either with the Cecchini "new system, 1900–05" or with the Mean Pole of Epoch 1962·0.

RÉSUMÉ

Il est nécessaire qu'on unifie la définition de la direction des petits axes des sphéroïdes utilisés pour les calculs géodésiques. En considération du mouvement du pôle, la définition "parallèle à l'axe de la rotation" est ambiguë. Peu de services géodésiques ont affirmé explicitement la direction adoptée, et ordinairement on l'a définie implicitement par les corrections pour le mouvement du pôle appliquées aux azimuts, latitudes et longitudes astronomiques. Mais on a négligé souvent ces corrections, ou on les a appliquées illogiquement, et il résulte des erreurs peut-être de $0''.5$ d'azimut. Le choix d'un Pôle Moyen Géodésique recommandé, par rapport auquel à l'avenir on peut prendre parallèles les petits axes des sphéroïdes, semble de reposer soit avec le Cecchini "new system, 1900–05", soit avec le Pôle Moyen de l'Epoque de 1962·0.

1. Notation

Geodetic Mean Pole	A direction, fixed in relation to the Earth, to which the minor axes of geodetic reference spheroids are to be parallel; denoted by a defined point on polar motion diagrams.
x_1, y_1	Coordinates of the instantaneous pole with reference to the Geodetic Mean Pole as in Figure 1 (seconds of arc).
x_2, y_2	Coordinates of the BIH mean pole of epoch with reference to the Geodetic Mean Pole (seconds of arc).
ϕ_G, λ_G, A_G	Geodetic latitude, longitude and azimuth at a field station.
ϕ_A, λ_A, A_A	Astronomical latitude, longitude and azimuth at a field station.
ϕ, λ	Approximate latitude and longitude at a field station.
ϕ_b	Astronomical latitude of a BIH observatory.
λ_b	Adopted astronomical longitude of a BIH observatory.

Markowitz and Guinot (eds.), Continental Drift, 37–43. © *I.A.U.*

w_b Weight of a BIH observatory. All latitudes positive north and longitudes positive east (astronomers use positive west).

LST_a Local sidereal time at a field station.

LST_b Local sidereal time at a BIH observatory.

R Greenwich apparent sidereal time at 0^h U.T.

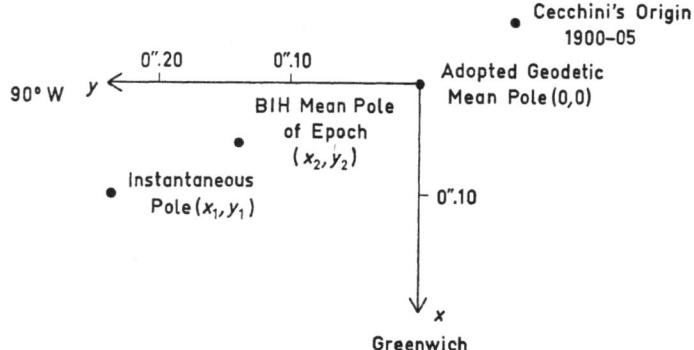

FIG. 1. *Note that the relative positions of the various poles are diagrammatic only.*

2. Geodetic Reference Systems

Geodesists compute positions in terms of latitude and longitude on a reference spheroid. For every detached survey they adopt arbitrary values for the major axis and flattening. They define the centre of the spheroid with reference to the direction of the vertical at some station known as their origin. Finally, they have to define the direction of the minor axis of the spheroid, and the direction of the zero meridian plane. In this paper we are only concerned with the last two items.

Loosely, the minor axis is always defined to be parallel to the Earth's axis of rotation. But if accepted geodetic latitudes and longitudes are to be constant, as they generally must be, the minor axis must be fixed in the rigid body of the Earth. Consequently, in view of the existence of Polar Motion, the minor axis of the spheroid cannot be defined as parallel to the instantaneous axis of rotation. The definition must either place it parallel to the line of points in the Earth which was the axis of rotation on some specified date, or else must refer to some mean position related to past directions of that axis.

The observed motion of the axis of rotation, or of the celestial pole, relative to the Earth, is customarily shown in a coordinate system such as that illustrated in Figure 1, which is of course fixed in relation to the Earth and shares its precession and nutation. Then the direction of the minor axis of the geodetic spheroid may be defined by designating any point on such a diagram as the adopted Geodetic Mean Pole.

When computations are carried out in Cartesian coordinates X, Y, Z, the direction of the Z-axis must be similarly defined, and the X-axis should lie in the Geodetic Zero Meridian, as defined in Section 5 below.

3. Formulae for Azimuth, Latitude and LST

At first sight a change in the direction of the minor axis, say θ'', would constitute a rotation which would change geodetic latitudes by comparable amounts, but in practice the geodetic latitude and longitude at the origin of any survey are held fixed (while the centre of the spheroid may be moved), so that the changes are much smaller, especially in a survey covering only a small area. The effect of changing the axis is correctly given if astronomical observations of latitude, longitude and azimuth are corrected as in equations (1)–(7) below.

An astronomical azimuth, observed with reference to the instantaneous axis of rotation, requires correction to the Geodetic Mean Pole as below.

Azimuth (mean) – Azimuth (Inst) =

$$-\left(x_1 \sin \lambda + y_1 \cos \lambda\right)'' \sec \phi. \tag{1}$$

Similarly the correction to an observed astronomical latitude is

$$+\left(y_1 \sin \lambda - x_1 \cos \lambda\right)''. \tag{2}$$

And the correction to an observed LST is

$$-\tfrac{1}{15}\left(x_1 \sin \lambda + y_1 \cos \lambda\right) \tan \phi \text{ seconds.} \tag{3}$$

4. Longitude and Laplace Azimuths

The correction to an observed astronomical longitude is complicated by the fact that U.T. 1 is reduced by the BIH to a variable Mean Pole of Epoch which in general cannot exactly coincide with the fixed Geodetic Mean Pole.

If all observations were reduced to the same pole we would have

$$\lambda_A = \mathrm{LST}_a - R - \mathrm{U.T.}\ 1$$
$$= \mathrm{LST}_a - R - \frac{1}{\Sigma w_b} \sum w_b (\mathrm{LST}_b - R - \lambda_b),$$

where the summation includes all observatories contributing to U.T. 1,

$$= \mathrm{LST}_a - \frac{1}{\Sigma w_b} \sum w_b (\mathrm{LST}_b - \lambda_b). \tag{4}$$

In equation (4), even though the field LST_a may have been corrected from instantaneous to some fixed pole by (3), the LST_b's will have been corrected by the BIH to the moving mean pole of epoch, and will each require the further correction

$$- \tfrac{1}{15}(x_2 \sin \lambda_b + y_2 \cos \lambda_b) \tan \phi_b. \tag{5}$$

So the total correction to a field longitude computed with U.T. 1 is

$$- \tfrac{1}{15}(x_1 \sin \lambda + y_1 \cos \lambda) \tan \phi$$

$$+ \frac{1}{15\Sigma w_b} \sum w_b (x_2 \sin \lambda_b + y_2 \cos \lambda_b) \tan \phi_b. \tag{6}$$

The geodetic Laplace azimuth at a field station is given by

$$A_G = A_A - (\lambda_A - \lambda_G) \sin \phi,$$

in which A_A requires correction as in (1), and λ_A as in (6), and the total correction to A_G is

$$- (x_1 \sin \lambda + y_1 \cos \lambda) \cos \phi$$

$$- \frac{\sin \phi}{\Sigma w_b} \sum w_b (x_2 \sin \lambda_b + y_2 \cos \lambda_b) \tan \phi_b. \tag{7}$$

For the 46 observatories used by the BIH in 1966 the value of $1/(\Sigma w_b) \sum (x_2 \sin \lambda_b + y_2 \cos \lambda_b) \tan \phi_b$ is

$$0 \cdot 29 \, x_2'' + 0 \cdot 42 \, y_2''. \tag{8}$$

The factor $0 \cdot 42$ is larger than might be expected, by reason of the uneven distribution of the observatories.

5. Zero Meridian Plane

The BIH zero meridian is a plane, containing the BIH Mean Pole of Epoch, through which the First Point of Aries passes at U.T. $1 = 24^h - R$. If all BIH observed LST's are corrected to the adopted Geodetic Mean Pole as in (5), the resulting Geodetic Zero Meridian will lie East of the BIH zero by

$$- \frac{1}{\Sigma w_b} \sum w_b (x_2 \sin \lambda_b + y_2 \cos \lambda_b) \tan \phi_b. \tag{9}$$

This zero meridian will be the zero of both geodetic and astronomical longitudes, λ_G and λ_A, as used by geodesists.

6. Magnitude of Errors

In equations (1)–(8) x_1 and y_1 vary about an annual mean position by not more than $0''.3$ in a year of large polar motion. If the adopted Geodetic Mean Pole is the Cecchini 1900–05 origin, y_1 may in the 1960's reach $0''.5$. The result of ignoring Equations (1)–(7) may then possibly be as much as $0''.5$, but generally not more than $0''.2$ or $0''.3$ ($\times \tan \phi$ for longitudes). Errors arising from the 12- or 14-month motion will be more or less random, but the difference between adopting one mean pole (e.g. 1962) and another (e.g. 1903) may be a systematic $0''.2$. The most serious consequence of neglecting the corrections is that Laplace azimuths may be wrong by these amounts. The corrections to astronomical latitudes and longitudes also cause changes in the deviations of the vertical, which affect geoidal sections and thence the reduction of measured distances to spheroid level, but in a survey extending less than (say) 4000 km from its origin this effect will be a few times less than that of the errors of azimuth.

7. Current Practice

In the past, few if any surveys have explicitly stated the defined direction of their minor axis, and there has been little or no consistency in the definitions implicitly made by reductions to Mean Pole. Some countries may have consistently applied equation (1) to observed astronomical azimuths, but it is more doubtful whether any have correctly reduced their Laplace azimuths. Possibly the nearest approach to consistency is when some country has used equations (1), (2) and (3) to reduce its observations to the BIH Mean Pole of Epoch. Then if its observations have not extended over many years, it can claim to have been nearly consistent, using the mean BIH mean pole over the period.

Writing in January 1967 we have not got full information about which countries may have been consistent, or nearly so.

8. Change of Spheroid

When detached surveys are computed with different spheroids and origins with (inevitably) different centres, a trustworthy common point makes it possible to reduce the latitudes and longitudes given by one into terms of the other. If the two surveys have parallel axes, the conversion is simple. If they have not, it is less simple.

9. Star Catalogues

Random probable errors, when brought up to the present epoch, are estimated to be of the order $0''.2$ for FK4, $2''$ for the Boss G.C. In both cases some stars, in particular those of high Southern declination, are subject to larger errors than are generally applicable, and in addition there will be considerable systematic error. Current observations should be computed with FK4, the catalogue on which the observations used by the BIH are based.

10. Future Needs

Errors such as those mentioned in Section 6 are less than the normal errors of observation: in Laplace azimuths about half the usual probable error or less. It is unlikely that any existing survey is materially the worse for their neglect.

It is to be expected, however, that satellite triangulation and other new techniques will increase accuracy, and now is the time to agree on a Mean Pole which should be adopted for future international geodetic work. We may hope that we are not acting too late, but little more delay can be afforded.

If any considerable geodetic computations had been consistently carried out using any particular mean pole, we would now be inclined to secure agreement to its adoption by all. Australia may be such a case. It is possible that there are others, but it is thought that there is unlikely to be any conflict between existing interests, and that we have a fairly free choice. What cannot be adopted is a continuously moving "Mean Pole of Epoch".

The choice seems to lie between two possible Mean Poles.

(a) The Cecchini "new system 1900–05". This is about 0″.2 distant from the present position of the mean. If there is any doubt about that figure, it would not be expedient to define the Geodetic Mean Pole as coinciding with the actual mean of 1900–05, but it could quite properly be defined as the origin of the IPMS polar coordinates. Then the current positions of the pole can be regarded as correct, while the geodetic mean pole is defined to be the zero point which gives them their accepted coordinates. The position of the pole 50 years ago might remain doubtful.

A great many individual observations have no doubt been reduced to this Cecchini origin, but it is doubtful whether any large survey, currently accepted and not soon due for recomputation, has been consistently computed in terms of it.

(b) The mean pole of epoch 1962·0. If a more recent position is to be accepted, this is a convenient date, since it coincides with the introduction of the FK4 catalogue and the new adopted longitudes in the work of the BIH. It avoids an unnecessary systematic correction of about 0″.2 in modern work, and it has just been accepted for Australia.

Subject to the comments of all those interested, it is thought likely that the mean pole of epoch 1962.0 will be found to be the most convenient definition of the Geodetic Mean Pole.

11. Action Now Required

Delegates at the Stresa symposium are asked to recommend a fixed Mean Pole for future geodetic reference systems.

12. Nomenclature

We have used the expressions "Geodetic Mean Pole" and "Geodetic Zero Meridian", as is necessary so long as other interests use a moving Pole. If a convenient fixed Pole and zero meridian could be adopted by all, there would be no need to define it as "Geodetic". It would suffice for geodesists to recommend its adoption for their work.

2.4. ON THE SECULAR VARIATION OF LATITUDE

T. OKUDA and C. SUGAWA

(International Latitude Observatory of Mizusawa, Japan)

ABSTRACT

The relation between the secular variation in latitude and the secular motion of the pole is discussed.

RÉSUMÉ

On discute la relation entre les variations séculaires des latitudes et le mouvement séculaire du pôle.

Viewing the distribution of the ILS stations, Mizusawa and Gaithersburg and also Kitab and Ukiah are nearly opposite to each other in longitude, in pairs, respectively. Mizusawa and Ukiah are situated nearly 90° in longitude apart from each other. Carloforte lies between these Mizusawa- and Ukiah-lines, and near the *x*-axis, or the meridian of Greenwich. We have taken these two lines as a frame of reference to re-examine the relation between the secular variation of latitude at the ILS station and the secular motion of the mean pole from the following two points of view:

(i) Comparisons of the secular variations of latitude on the same line such as Mizusawa–Gaithersburg and Ukiah–Tschardjui (or Kitab).

(ii) Projection of the resultant vector composed of the relative velocities of the secular variations in latitude along Mizusawa- and Ukiah-lines on the meridian at Carloforte.

Dividing the whole period 1900–66 into five intervals, each of which covers about 12 years, we have put the above tests for each interval. Consequently we may conclude that a major part of the secular variation in latitude at any station would be caused by the secular motion of the mean pole, but the remaining part would be attributed to the crustal movement of the station or the local change in the plumb line at the station. Moreover, we have noticed the fact that conspicuous local non-polar variations along Mizusawa-line give opposite phase to those along Ukiah-line with about 19 years' period. The details about these variations will be published in the near future.

Comparing the results of latitude observations made at Tokyo and Tientsin with those made at Mizusawa for the recent period, Orlov mean latitudes at these stations give fairly similar variations for the period 1959–63. Therefore, no noticeable relative motion seems to exist between Japan islands and Asian continent. We may consider at present that Japan islands tend to move regionally together with Asian continent.

Markowitz and Guinot (eds.), Continental Drift, 44. © *I.A.U.*

3.1. SECULAR VARIATION OF LONGITUDE AND RELATED PROBLEMS

M. Torao, S. Okazaki, and S. Fujii

(Tokyo Astronomical Observatory, Japan)

ABSTRACT

The secular variations in longitude differences among several observatories were determined, but it is hard to consider that these results are literally due to continental drift. The large standard deviations of the data of longitude differences show the existence of external systematic errors in time observations.

RÉSUMÉ

Des variations séculaires des différences de longitudes entre plusieurs observatoires ont été trouvées; mais il est difficile de les considérer comme réellement dues à des dérives continentales. Les grands écarts-types sur les différences de longitudes montrent l'existence d'erreurs systématiques externes dans les observations de l'heure.

1. Introduction

We have analyzed the time observations of various observatories and have derived relative variations of longitude. The observatories and their combinations are designated as follows:

G	Greenwich	W	Washington
H	Hamburg Hydrographic	R	Richmond
Pa	Paris	BAg	Buenos-Aires Geodetic
Pt	Potsdam Geodetic	BAn	Buenos-Aires Naval
U	Uccle	To	Tokyo
O	Ottawa	Mz	Mizusawa

E_5 mean of G, H, Pa, Pt, and U.
W_2 mean of W and R.
BA_2 mean of BAg and BAn.

The effect of polar motion was removed by using the coordinates of the instantaneous pole referred to a fixed pole, the mean pole from 1900·0 to 1906·0. Longitudes are positive to the west.

2. Tendency for Long-Period Variations

The summarized results from time observations for 30 years, 1933–62, for several combinations of observatories are shown in Table 1 and Figure 1.

Markowitz and Guinot (eds.), Continental Drift, 45–51. © *I.A.U.*

Table 1

Secular variations of longitude, in ms/yr, from 30 years data

	Sec. Var.	m.e.
To–W_2	$+1{\cdot}05$	$\pm 0{\cdot}19$
To–BAg	$+1{\cdot}51$	$\cdot35$
To–E_5	$+0{\cdot}04$	$\cdot19$
W_2–E_5	$-1{\cdot}26$	$\cdot17$
BAg–E_5	$-1{\cdot}11$	$\cdot29$
W_2–BAg	$-0{\cdot}15$	$\cdot33$

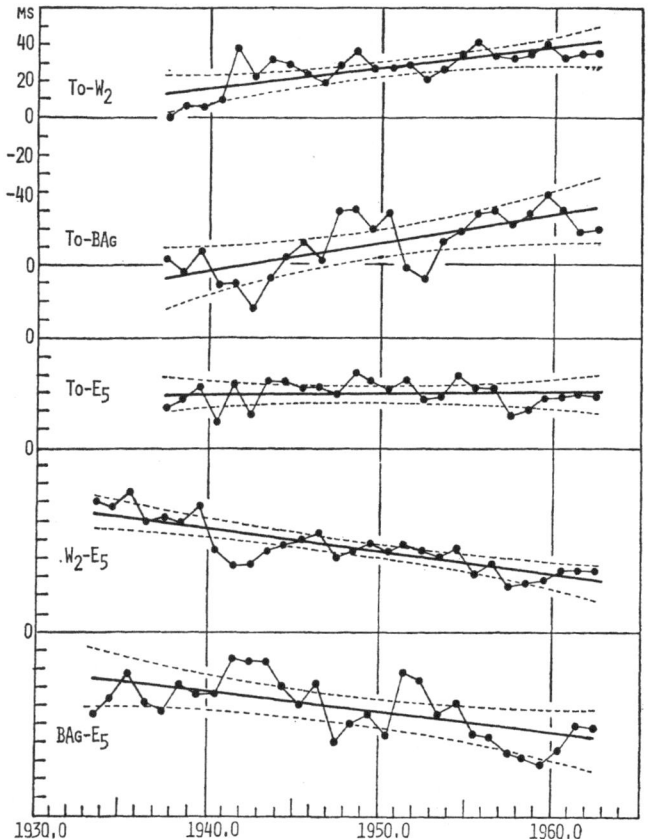

FIG. 1. *Secular variation of the difference in longitude; result (I). Vertical ordinate shows relative difference in longitude.*

3. Solution for Short Intervals

When we divided the data into two parts, (a) and (b), before and after 1950·0, different results were obtained, as shown in Table 2 and Figure 2.

Table 2

Secular variations of longitude for short intervals

	(a) 1933–49		(b) 1950–62		(c) 1956.8–1965.9	
To–W₂ †	+2·26	±0·63	+0·81	±0·39	−1·64	±0·24
To–BAg †	+2·61	·97	+1·65	1·01	−0·39	·45*
To–E₅ †	+1·16	·49	−0·57	·51	−0·48	·35
W₂–E₅	−1·76	·41	−1·38	·48	+1·23	·29
BAg–E₅	−0·46	·62	−2·22	1·06	−0·09	·49*
W₂–BAg	−1·30	·86	+0·84	·83	+1·32	·46*

* BA₂.   † 1937–49.

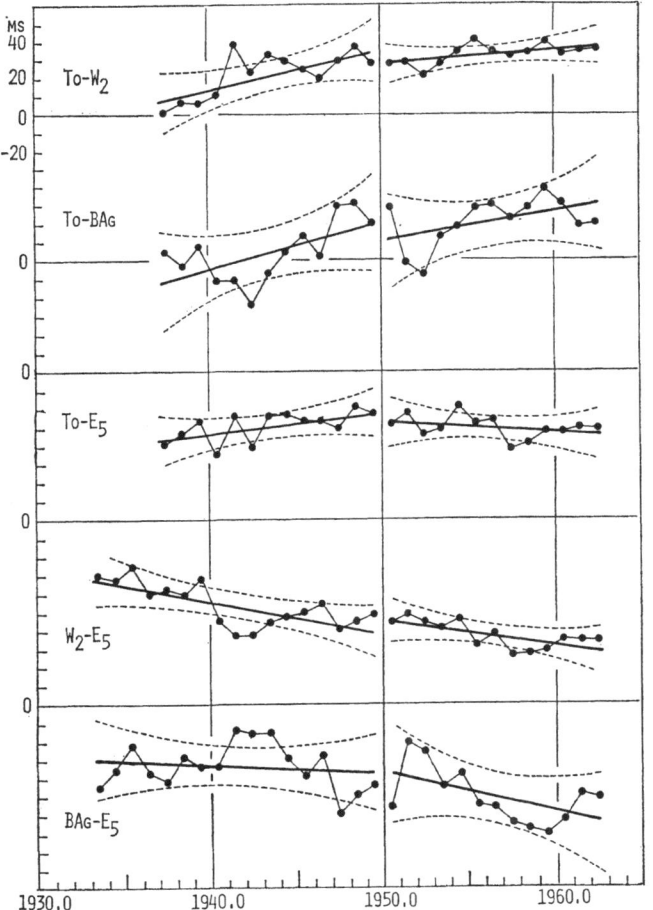

FIG. 2. *Secular variation of the difference in longitude; results (a) and (b).*

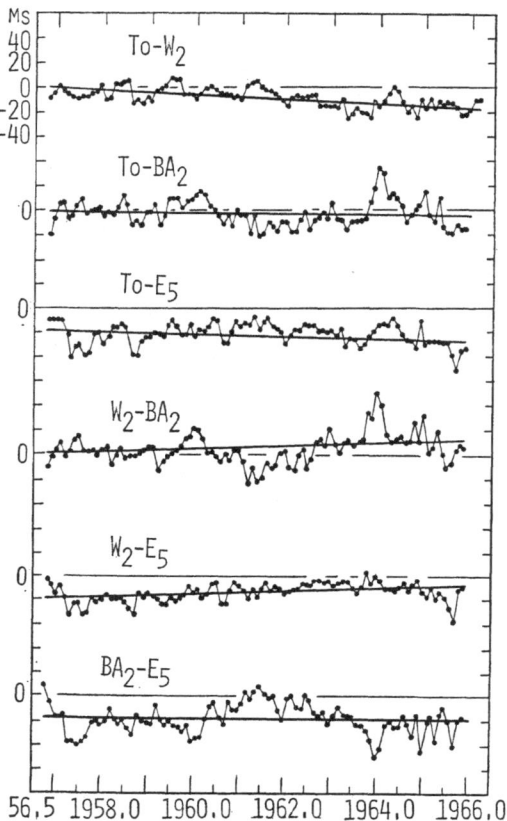

FIG. 3. *Secular variation of the difference in longitude; result (c).*

Table 3

Secular variations of longitude between nearby observatories, from recent data, 1956·8–1965·5

	Sec. Var.	m.e.
G–Pa	−1·19	±0·41
G–H	+0·57*	·58
G–Pt	−1·37	·48
G–U	+3·45	·73
G–E₅	+0·32	·29
W–Rc	+0·27	·33
W₂–O	+1·73*	·41
To–Mz	−0.07**	·47
BAn–BAg	−2·60	·66

* After 1960·5.
**After 1958·1.

We can use the atomic time scale from the end of 1956, so we reduced the results in terms of A.1, which is maintained by the Naval Observatory, Washington. We determined the secular variations of longitude differences for the recent 11-year interval, 1956·8–1965·9. The results are shown as (c) in Table 2 and Figure 3. There are large discrepancies between the results (a), (b), (c), and those obtained in Section 2.

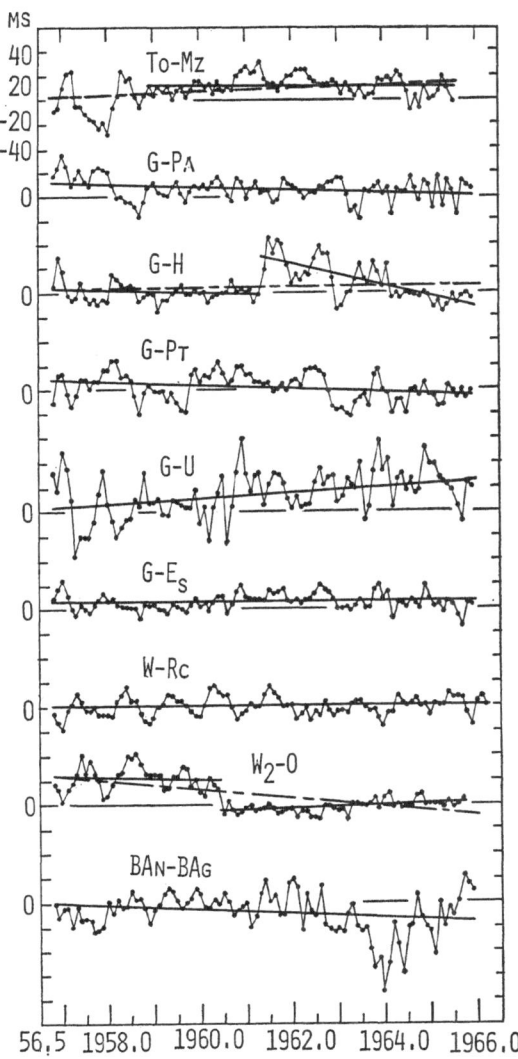

FIG. 4. *Secular variation of the difference in longitude between nearby observatories.*

Table 4

Variations in longitude relative to mean observatory and standard deviations, 1956·8–1965·5

	a(ms)		b(ms/y)		s·d·(ms)
To	− 8·9	±0.7	−0·61	±0·24	± 5·7
Mz	−16·1	1.1	−1·96	·38	9·1
			(−0·54)*		
G	+17·1	0.6	−0·33	·22	5·2
Pa	+ 9·5	1.0	+0·86	·36	8·7
H	+12·7	1.5	−0·91	·52	12·5
Pt	+14·0	1.1	+1·03	·40	9·6
U	+ 6·8	1.8	−3·78	·65	15·4
W	− 3·5	0.5	+1·12	·18	4·4
Rc	− 3·2	0.8	+0·86	·29	6·9
O	−15·8	1.2	+4·86	·42	9·9
			(−0·61)**		
BAg	− 3·4	1.0	+0·73	·37	8·7
BAn	− 9·9	1.7	−1·87	·63	14·8

* After 1958·1.
** After 1960·0.

4. Longitude Differences between Nearby Observatories

Using recent time observations, the secular variations of longitudes between several pairs of nearby observatories were determined. The results are shown in Table 3 and Figure 4.

5. Separate Results for Each Observatory

From the recent data (c) in Section 3, we determined the standard deviation of a longitude determination relative to the mean longitude of 12 observatories for each 0.1 year interval, using the residuals:

$$r_i = [\text{U.T. } 1_i - \text{U.T. } 1_m] - [a_i + b_i(t - 1960·0)],$$

where U.T. 1_m is the mean U.T. 1_i. The results are given in Table 4.

6. Discussion and Conclusions

The secular variations of longitude differences for various combinations of observatories as derived from long intervals seem to be significant, considering their mean errors. However, when we divide the data into two periods, very different results occur. The same situation appears for results from the recent, precise time observations.

Regarding these circumstances, it is hard to consider that the observed secular variations of longitudes are due to continental drift. The large variations appearing for some pairs of nearby observatories lead to the same conclusion.

The external systematic error may be fairly large in the time observations. Because of this, the standard deviations shown in Section 5 are larger than those estimated from the internal accuracies of the recent observations. We have used in this study data published formally by each observatory. However, systematic effects introduced by changes in instruments or locations and by different star catalogues may not have been completely removed. The residual effects may be fairly large, as shown by Table 4 and Figure 4.

Unification of the systems of star places and proper motions of time and latitude stars is urgently needed. Studies of the effects of meteorological conditions are also important.

With the present accuracies, as shown in Section 3, about 20 or more years are needed for determining the secular variations of longitude within an accuracy of ± 0.1 ms/yr.

The general conclusion of the present paper is that there have been fairly large systematic errors in time observation so that the apparent secular changes in longitudes cannot be due to continental drift literally.

3.2. MOUVEMENT SÉCULAIRE DU PÔLE ET LA VARIATION DES LATITUDES DES STATIONS DU SIL

ANNA STOYKO

(Observatoire de Paris, France)

RÉSUMÉ

Les résultats du SIL réduits dans un système uniforme donnent pour la vitesse moyenne annuelle du mouvement du pôle la valeur de 0″0032 et pour sa direction celle de 70°W. La latitude de Mizusawa a diminué de 0″003 par an et celle d'Ukiah a augmenté de la même quantité par an an pendant la période étudiée. Cela confirme la théorie géologique indiquant la rotation des côtes du Pacifique dans le sens contre celui des aiguilles d'une montre.

РЕЗЮМЕ

Результаты МСШ, приведенные к одной общей системе, дают для средней годовой скорости движения полюса значение, равное 0″0032, а для его направления 70°W. Широта Мицузавы уменьшилась на 0″003 в год, а широта Юкайи увеличилась на то же количество в год за рассматриваемый период времени. Это подтверждает геологическую теорию, показывающую вращение берегов Тихого океана в направлении, противоположном направлению движения часовой стрелки.

ABSTRACT

The ILS results reduced to a uniform system give a motion of the pole of 0″0032/year in the direction 70°W. The latitude of Mizusawa decreases 0″003/yr and that of Ukiah increases at the same rate. This confirms the geological theory which indicates a counter-clockwise rotation of the coasts of the Pacific.

Les nouvelles valeurs de latitudes conventionnelles des stations du SIL, introduites par Cecchini en 1959, et l'application des corrections que nous publions dans le Tableau 1 permettent de réduire les coordonnées du pôle instantané déjà publiées à un système uniforme. Pour mieux uniformiser ces coordonnées, nous les avons recalculées depuis 1900 en traitant les résultats des stations du SIL de l'hémisphère Nord. Préalablement nous avons affranchi les latitudes observées des erreurs de catalogue par la méthode en chaîne quand cela était nécessaire.

Nous avons lissé les coordonnées du pôle instantané et nous leur avons appliqué la méthode d'Orlov. Leurs moyennes annuelles sont représentées sur la Figure 1 avec l'adjonction de la période de la collaboration libre de 1890 à 1899. Afin d'éliminer les irrégularités accidentelles des valeurs annuelles dont témoigne la Figure 1, nous avons pris les moyennes glissantes de 6 en 6 ans, représentées sur la Figure 2. La périodicité à longue durée n'est pas apparente, mais on remarque des changements assez brusques

Tableau 1

Réductions des coordonnées x et y du pôle instantané et du terme non polaire z au nouveau système de Cecchini

(Unité: millième de seconde d'arc)

Période	Δx	Δy	Δz	Période	Δx	Δy	Δz
1899·9–1912·0	− 8	−13	− 3	1935·0–1940·7	+29	+139	−28
1912·0–1915·0	− 3	+22	+ 10	1940·8–1941·0	+49	+138	−21
1915·0–1916·0	− 3	+21	+ 10	1941·0–1941·5	+ 9	+ 77	−30
1916·0–1919·4	− 5	+16	+ 7	1941·5–1943·2	+34	+ 77	−20
1919·5–1922·7	− 5	+10	+ 6	1943·3–1946·4	+ 9	+ 77	−30
1922·7–1930·8	+43	+89	−116	1946·4–1949·0	+34	+ 77	−20
1930·9–1932·6	+41	+67	−116	1949·0 →	+48	+ 43	−35
1932·7–1935·0	+39	+88	−121				

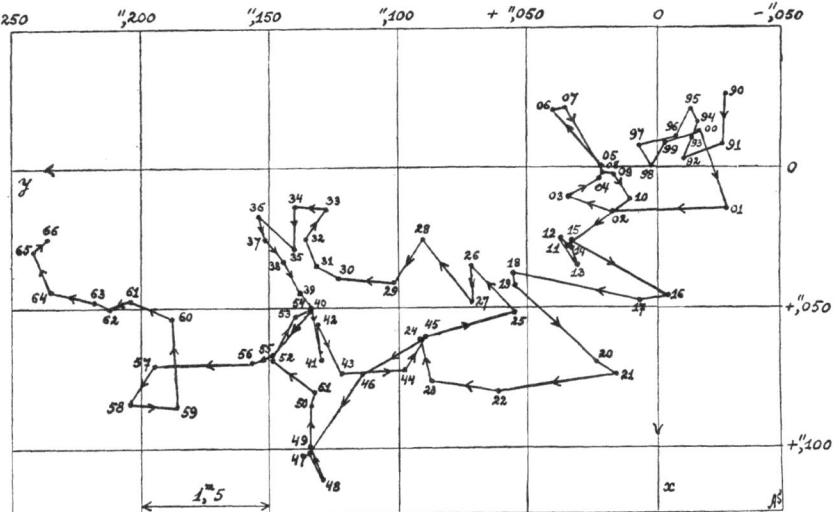

Fig. 1. *Mouvement du pôle d'après toutes les stations nord du SIL (système uniforme, moyennes annuelles par la méthode d'Orlov).*

dans la direction du mouvement du pôle, ne coïncidant pas toujours avec les modifications de programme d'observations.

D'après les valeurs de la Figure 2, nous avons calculé la vitesse moyenne annuelle (Va) et la direction (D) du mouvement du pôle instantané pour la période de 77 années: $Va = 0''.0032$, $D = 69°8$ W. Ensuite, d'après les résultats des trois stations du SIL (Mizusawa, Carloforte et Ukiah) qui ont travaillé d'une façon presque continue à partir de 1900, nous avons trouvé pour la vitesse et la direction du mouvement séculaire du pôle les valeurs: $Va = 0''.0040$ et $D = 73°0$ W, qui sont plus fortes que celles qui ont été déterminées d'après l'ensemble de toutes les stations nord du SIL.

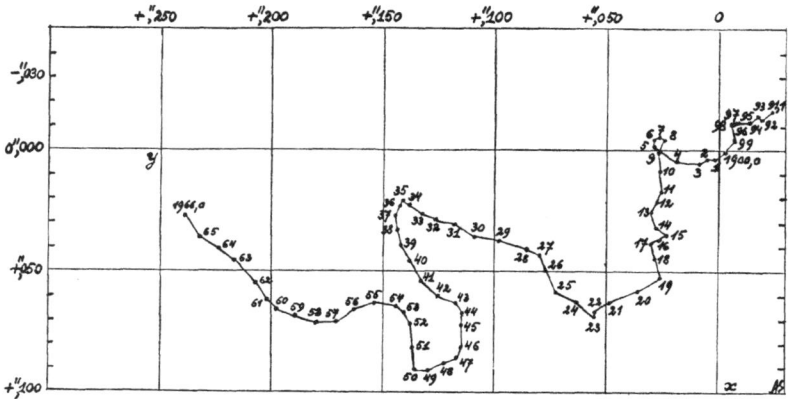

FIG. 2. *Mouvement séculaire du pôle d'après les stations nord du SIL (moyennes glissantes de 6 en 6 années).*

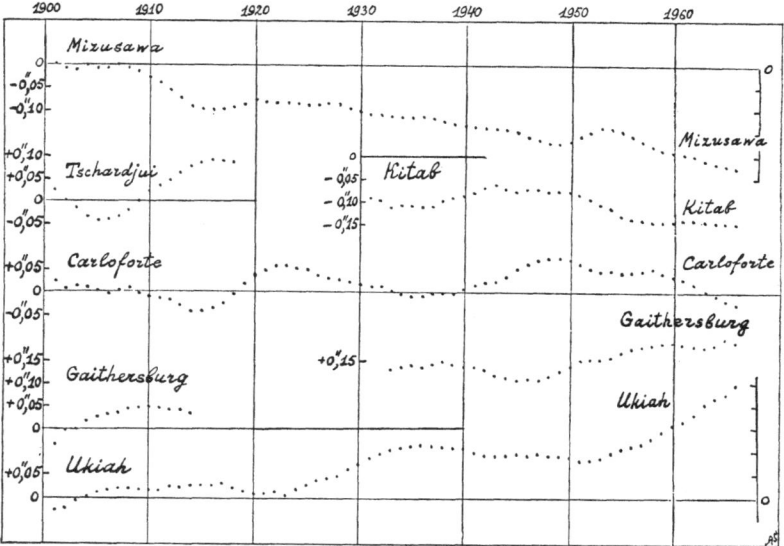

FIG. 3. *Variation séculaire des latitudes des stations du SIL (moyennes glissantes de 6 en 6 années).*

La détermination de la variation séculaire des latitudes présente des difficultés plus importantes que celle de la variation séculaire du mouvement du pôle instantané. Soient φ_i et φ_{ci} les latitudes observée et conventionnelle d'une station i et x_1 et y_1 les coordonnées du pôle instantané par rapport au pôle moyen de l'époque. Dans ce cas la correction de latitude conventionnelle est égale à

$$\Delta\varphi_i = \varphi_i - \varphi_{ci} - (x_1 \cos \lambda_i + y_1 \sin \lambda_i) - z.$$

Les valeurs $\Delta\varphi_i$ sont indépendantes des erreurs du catalogue d'étoiles et des erreurs systématiques non polaires communes à toutes les stations du SIL. Les moyennes glissantes de 6 en 6 ans, donnant la variation séculaire des latitudes, sont représentées graphiquement sur la Figure 3. D'après ces résultats nous avons trouvé les vitesses moyennes annuelles de latitudes par les moindres carrés (M.C.) et par la méthode graphique (Gr.). La partie gauche du Tableau 2 contient ces valeurs et leurs moyennes (Moy.). Les deux stations de l'Amérique du Nord donnent la même valeur de la vitesse moyenne annuelle, correspondant à l'augmentation de latitudes de ces stations de 0″003 par an (9 cm par an).

Ce résultat suggère l'idée que les Etats-Unis d'Amérique se déplaceraient en un seul bloc vers le nord. Par contre, la station de Mizusawa descendrait dans la direction de l'équateur. Ce phénomène serait en bon accord avec la théorie géologique, selon laquelle les côtes du Pacifique ont un mouvement de rotation dans le sens contraire à celui de l'aiguille d'une montre. En réduisant ces vitesses au pôle fixe, on trouve les variations annuelles de latitudes représentées dans la partie droite du Tableau 2. On remarque une diminution de vitesses séculaires de latitudes en valeur absolue.

Mme Obrezkova (1966) a déterminé la variation des latitudes des stations Mizusawa Carloforte et Ukiah pour la période de 1901 à 1934, en réduisant les résultats d'observations de leurs latitudes aux trois catalogues différents: F. Cohn (F.C.), N30 et GC. Dans le Tableau 3 nous donnons les vitesses moyennes annuelles de latitudes

Tableau 2

Mouvement moyen annuel progressif des latitudes des stations nord du SIL

(Unité: millième de seconde d'arc)

	Réduction au pôle moyen de l'époque			Réduction au pôle fixe (latitudes constantes)		
	M.C.	Gr	Moy	M.C.	Gr	Moy
Mz	−3,01	−3,20	−3,10	−0,26	−0,54	−0,40
Kt	−1,78	−1,34	−1,56	+0,42	+1,03	+0,72
Car	+0,59	0,00	+0,29	−0,07	−0,37	−0,22
Gaith	+2,94	+3,06	+3,00	−0,23	−0,60	−0,42
Uk	+2,90	+3,25	+3,07	+1,00	+0,89	+0,95

Tableau 3

Mouvement moyen annuel progressif des latitudes de 1901 à 1934

(Unité: millième de seconde d'arc)

	AS	FC	N30	GC
Mizusawa	−3,46	−1,30	−3,61	−5,94
Carloforte	0,00	+2,18	−0,30	−2,82
Ukiah	+3,39	+5,39	+3,09	+0,91

calculées d'après les résultats d'Obrezkova et, à titre de comparaison, les vitesses obtenues d'après nos calculs (AS) pour la même période, indépendantes d'erreurs des mouvements propres de catalogue. Les valeurs AS se rapprochent le plus des résultats calculés avec le catalogue d'étoiles N30 (Tableau 3). On remarque d'autre part que le catalogue FC donne pour Mizusawa la vitesse de la variation de latitude la plus faible en valeur absolue et la plus forte pour Ukiah, tandis que le catalogue GC donne le résultat opposé.

Ainsi, la nécessité d'avoir un catalogue commun pour toutes les étoiles de latitude apparaît clairement. Il est donc désirable que l'élaboration d'un tel catalogue, déjà commencée, soit terminée le plus rapidement possible.

Référence

Obrezkova, E. I. (1966)   *Geofis. and Astr. Inform. Bull.* Kiev, **9**, 145. (En russe.)

3.3. VARIATION SÉCULAIRE DES LONGITUDES

Nicolas Stoyko

(Observatoire de Paris, France)

RÉSUMÉ

La réduction des résultats de services horaires dans un système uniforme permet de déterminer la variation séculaire des longitudes avec une grande précision. Il faut considérer un long intervalle de temps pour supprimer l'influence des variations quasi-périodiques dans la détermination des longitudes. On remarque des variations concomitantes des longitudes des stations rapprochées, se trouvant sur un même bloc de l'écorce terrestre, ou des variations de sens opposé, quand les stations étudiées se trouvent sur des blocs voisins, se déplaçant l'un par rapport à l'autre.

РЕЗЮМЕ

Приведение результатов служб времени к одной системе позволяет определить вековое изменение долгот с большой точностью. Необходимо использовать большой промежуток времени, чтобы уничтожить влияние квази-периодических вариаций в определении долгот. Замечено одновременное изменение долгот близлежащих станций, находящихся на одном блоке земной коры, а также движения в противоположных направлениях станций, находящихся на соседних блоках, которые перемещаются один по отношению к другому.

ABSTRACT

The reduction of time service results on a uniform system permits the secular determination of longitudes with high precision. A long interval of time is needed to eliminate quasi-periodic effects. One notes that the variations are similar when stations are in the same terrestrial region and opposed when in different regions.

Nous avons révisé les résultats de services horaires de 1925 à 1965 en prenant en considération les nouvelles valeurs de longitudes conventionnelles et le catalogue FK4. Nous avons ramené ces résultats à l'Observatoire moyen composé de 11 services horaires ayant travaillé pendant toute la période étudiée. Sur les Figures 1 et 2 nous avons représenté les courbes des corrections des longitudes ainsi calculées pour chaque observatoire. Nous avons déterminé d'après ces valeurs les corrections de longitudes pour l'année 1965 ($\Delta\lambda_{65}$) et leurs variations annuelles (Δm_a) qui figurent dans le Tableau 1. En plus ce tableau contient la variation annuelle des longitudes déterminée d'après la méthode graphique (Δm_{ag}).

Comme les variations de longitudes ne sont pas tout-à-fait linéaires, nous publions dans le Tableau 2 les marches annuelles de longitudes de 10 en 10 ans et leurs écarts moyens (E*m*). La dernière colonne (M_{40}) contient la moyenne de 40 années. On voit que les variations annuelles de longitudes pour les différentes périodes sont très

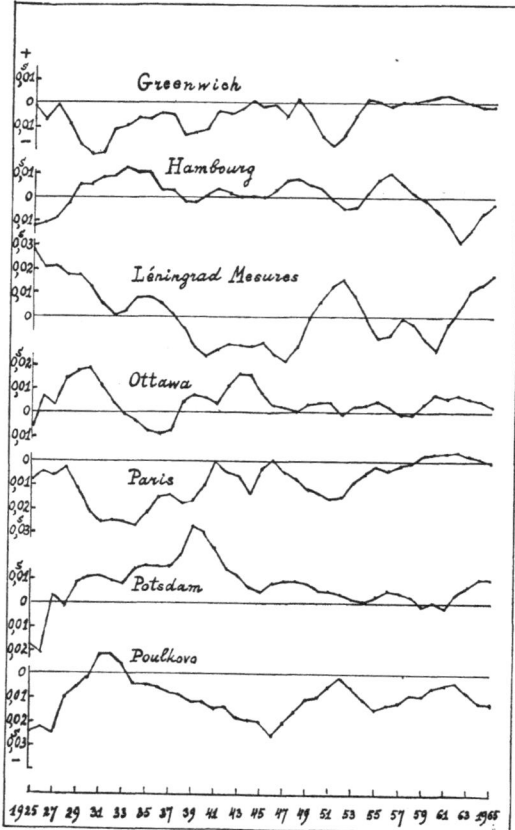

FIG. 1.   *Variations séculaires des longitudes.*

discordantes (de l'ordre de 0".0006), ce qui explique les conclusions quelquefois contra-dictoires.

La variation annuelle de longitudes entre l'Amérique du Nord et l'Europe pour la période de 1925 à 1965 est de −0".00060 (rapprochement des continents). Pour le contrôle nous avons traité les résultats de la détermination des longitudes à partir de 1871, réduits au même système et n'entrant pas dans le service courant de l'heure (Tableau 3). Ces résultats, couvrant la période de 88 ans, donnent la marche annuelle de −0".00068, ce qui est en bon accord avec la valeur précédente.

Selon Wegener il devrait exister un déplacement important de Madagascar par rapport à l'Afrique du Sud (éloignement) et celui de Groenland vers l'Ouest. D'après Stoyko (1966) la variation annuelle de la longitude de Madagascar vers l'Est est de −0".00033.

D'après les deux premières opérations des longitudes mondiales le déplacement

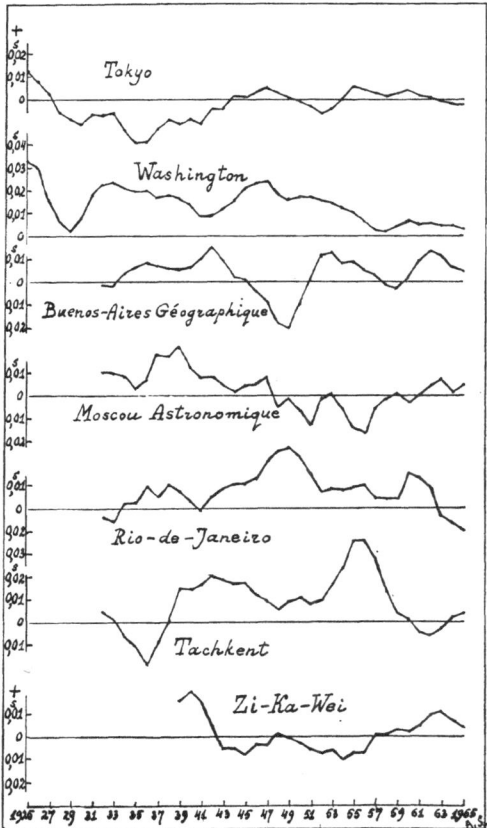

FIG. 2. *Variations séculaires des longitudes.*

annuel du Cap vers l'Est est égal à −0″.00148, ce qui correspond à la diminution de la distance entre l'Afrique du Sud et le Madagascar (43 cm par an).

La détermination moderne de la longitude de la station de Kornok (Groenland) a été faite en 1927 (Sindig, 1955). Les observations de contrôle, faites en 1948, ont donné une valeur supérieure de 0″.02 par rapport à celle de 1927, ce qui indique un déplacement faible de Groenland vers l'Ouest (19 cm par an).

Les mouvements récents de l'écorce terrestre se produisent par blocs qui peuvent avoir d'assez vastes dimensions. Dans ce cas les stations qui se trouvent sur un même bloc doivent avoir les variations concommitantes de longitudes. Nous avons étudié dans ce but le mouvement des couples suivants de stations: Greenwich–Paris, Hambourg–Potsdam, Léningrad–Poulkovo et Ottawa–Washington.

En éliminant les variations séculaires de longitudes de Greenwich et de Paris d'après le Tableau 1, nous avons trouvé que les variations résiduelles de longitudes à courte

Tableau 1

Corrections des longitudes pour l'année 1965 et leurs variations annuelles par rapport à l'Observatoire moyen constant

(En millièmes de seconde de temps)

			Moindres carrés				Graph.
Noms	Abr^n	n	$\Delta\lambda_{65}$	Em	Δm_a	Em_a	Δm_{ag}
Buenos Aires g.	BAg	34	+ 4·4	±3·7	+0·05	±0.14	+0·10
Greenwich	Gr	41	+ 0·5	1·7	+0·30	7	+0·25
Hambourg	H	41	− 2·4	2·2	−0·16	9	−0·25
Léningrad mes.	Lm	41	− 4·5	3·7	−0·29	16	−0·32
Moscou astr.	Ma	34	− 5·1	3·3	−0·49	13	−0·38
Neuchâtel	N	41	− 2·1	5·2	−0·97	22	"
Ottawa	O	41	+ 4·1	1·9	−0·03	8	−0·27
Paris	Pa	41	+ 1·0	2·1	+0·46	9	+0·40
Potsdam	Pt	41	+ 5·9	2·7	−0·09	12	−0·05
Poulkovo	Pu	41	− 9·8	2·4	+0·01	11	−0·05
Rio de Janeiro	RJ	34	+ 8·4	3·8	+0·03	15	+0·10
Tashkent	Ta	34	+12·5	5·5	+0·23	22	+0·37
Tokyo	To	41	+ 1·9	2·0	+0·22	9	+0·12
Uccle	U	41	− 5·4	7·9	+0·88	34	"
Washington	W	41	+ 6·2	1·9	−0·40	8	−0·52
Zi-Ka-Wei	Zi	27	+ 0·2	5·7	−0·06	20	−0·20

Tableau 2

Variations annuelles des longitudes de 10 en 10 ans et leur moyenne

(En millièmes de seconde de temps)

	Marche annuelle					
Abr^n	1926 1935	1936 1945	1946 1955	1956 1965	Em	M_{40}
BAg	"	−0·31	+2·97	+0·79	±0·54	+1·15
Gr	−0·10	+0·88	0·00	−0·09	·18	+0·17
H	+2·51	−0·16	−0·55	−2·04	·72	−0·06
Lm	−2·03	−1·69	+2·11	+2·67	·76	+0·27
Ma	"	−1·50	−1·56	+1·80	·44	−0·42
N	+3·29	−3·09	−2·01	−0·69	·78	−0·63
O	−1·78	+2·29	+0·13	+0·51	·29	+0·29
Pa	−2·76	+1·50	−0·32	+0·42	·24	−0·29
Pt	+2·77	−1·57	−1·00	+0·76	·42	+0·24
Pu	+2·59	−1·66	+1·56	+0·20	·39	+0·67
RJ	"	+0·13	−1·65	−1·70	·78	−0·74
Ta	"	+3·67	+2·22	−3·52	1·05	+0·79
To	−2·09	+1·97	−0·43	−0·70	·13	−0·31
U	−2·91	+0·81	+1·49	−2·67	·35	−0·82
W	+1·17	−0·33	−1·29	−0·03	·16	−0·12
Zi	"	−5·04	−0·85	+1·35	±0·37	−1·51

Tableau 3

Corrections des différences de longitudes entre l'Amérique du Nord et l'Europe occidentale réduites à un même système et aux nouvelles longitudes conventionnelles

	W–Eu		W–Eu
1871	$+0''0630$	1926	$+0''0095$
1892	$+$ 450	1933	$+$ 193
1914	$+$ 180	1959	$+$ 030
1922	$+$ 179		

Tableau 4

Variations annuelles des longitudes entre les continents d'après les longitudes mondiales et les services horaires permanents

	Long. mond.	Service perm.
Période	1930,3–1958,8	1925–1965
W–Eu4	$-0''00030$	$-0''00055$
Eu4–To	$-$ 28	$-$ 07
To–W	$+$ 58	$+$ 62

durée sont, en majorité de cas, du même sens (75%). Le coefficient de corrélation entre leurs mouvements est $r = +0·6$. Comme ces deux observatoires sont complétement indépendants l'un de l'autre, on peut supposer qu'ils se trouvent sur un même bloc de l'écorce terrestre. La même conclusion est valable pour les autres couples de l'Europe.

Dans le cas des observatoires d'Ottawa et de Washington on remarque qu'après l'élimination du terme séculaire, leurs mouvements à courte période sont de sens opposé. Le coefficient de corrélation est $r = -0·6$. Cela peut provenir du fait que ces observatoires se trouvent sur les deux blocs différents de l'écorce terrestre, dont la ligne de séparation est dirigée dans le sens WS–EN (Grands lacs, fleuve St. Laurent). Si les longitudes du bloc Sud diminuent (mouvement vers l'Est), les latitudes doivent augmenter en raison du glissement de deux blocs le long de la ligne de la séparation. Les résultats de la station de latitude de Gaithersburg, indépendants du catalogue d'étoiles, montrent d'après l'étude de Stoyko (1967) l'existence du déplacement de cette station vers le Nord de $0''003$ par an, ce qui est en accord avec la variation de longitude de Washington vers l'Est. Nous avons trouvé un coefficient de corrélation très fort $r = -0·92$ entre la variation de longitude de Washington et de latitude de Gaithersburg.

Dans le Tableau 4 nous donnons les conclusions sur le mouvement séculaire des continents d'après les résultats des trois opérations des longitudes mondiales (Stoyko, 1966). Les résultats obtenus montrent l'existence d'un rapprochement entre l'Amérique du Nord et l'Europe, ainsi qu'entre l'Europe et le Japon et d'un éloignement entre le Japon et l'Amérique du Nord (élargissement de l'Océan Pacifique). Dans la partie droite du Tableau 4 nous donnons les résultats obtenus d'après les services horaires permanents, confirmant la conclusion précedente.

Références

Sindig, E. (1955)   *Mém. Inst. Géodés. Danemark*, **19**, 19.
Stoyko, Mme A. (1966)   *Ann. Intern. Geophys. Year*, **42**, 1–369.
Stoyko, Mme A. (1968)   the present volume, p. 52.

4.1. NOUVELLES MÉTHODES DE CALCUL DU BUREAU INTERNATIONAL DE L'HEURE

BERNARD GUINOT et MARTINE FEISSEL

(Observatoire de Paris, France)

RÉSUMÉ

En développant de nouvelles méthodes de calcul pour le BIH, nous sommes arrivés aux conclusions suivantes: (1) il est nécessaire d'exprimer les résultats du BIH en utilisant une origine fixe pour les coordonnées du pôle; (2) la mesure des dérives continentales à latitudes égales doit être faite par des PZT; à des latitudes très différentes la contribution des astrolabes est utile.

ABSTRACT

In developing new methods of computation for the BIH we reached the following conclusions: (1) it is necessary for the BIH to use polar coordinates referred to a fixed origin; (2) the measurement of continental drifts by stations on nearly the same latitude may be carried out with PZT's. When the latitudes are very different, astrolabes can provide useful results.

Afin de publier rapidement l'heure définitive du BIH et éviter que les lissages conduisant aux heures demi-définitives des stations participantes ne fassent disparaître des détails de la rotation de la terre, nous avons cherché à utiliser les mesures brutes de TU0, en combinaison ou non avec les mesures de latitude, pour en déduire TU1 et les coordonnées x et y du pôle. Nous donnons ici quelques résultats des travaux en cours.

Le système d'équations à résoudre est:

$$x \cos L_0 + y \sin L_0 \, (+ z) = \varphi - \varphi_0, \tag{1}$$

$$- x \operatorname{tg} \varphi_0 \sin L_0 + y \operatorname{tg} \varphi_0 \cos L_0 + t = \mathrm{TU0} - \mathrm{TUC} + \varDelta Ts, \tag{2}$$

où φ_0 et L_0 sont des valeurs fixes des latitude et longitude de référence (longitude du BIH pour L_0); φ, la valeur observée instantanée de la latitude; TU0−TUC, la valeur observée de la différence entre TU0, calculé avec L_0, et TUC, le temps coordonné transmis par les signaux horaires; $\varDelta Ts$, la correction conventionnelle du BIH pour l'irrégularité saisonnière de la rotation de la terre, introduite pour des raisons de commodité.

L'inconnue t est TU2−TUC. Les solutions ont été essayées avec et sans l'inconnue z (terme de Kimura).

Dans cette première phase du travail, les valeurs observées, telles qu'elles sont

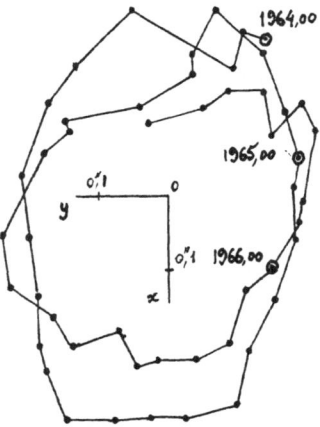

FIG. 1. *Polhodie d'après les mesures de temps. 8 PZT et 8 astrolabes.*

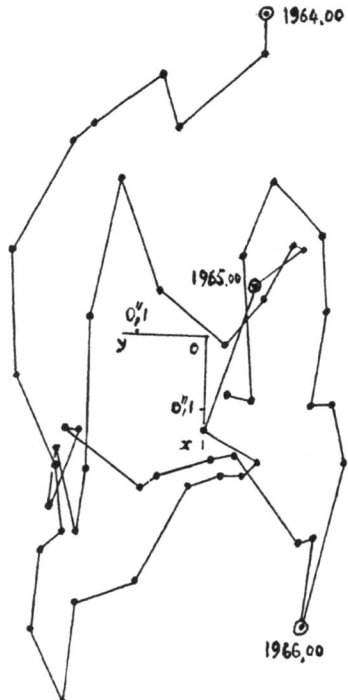

FIG. 2. *Polhodie d'après les mesures de temps. 25 instruments des passages.*

transmises par les stations, sont groupées par moyennes de poids égaux, les poids de chaque moyenne étant choisis de sorte qu'on ait environ 20 points de mesure par an. La solution est ajustée par la méthode des moindres carrés, pour chaque $\frac{1}{20}$ d'année. Le second membre des Équations (1) et (2) est, en principe, interpolé linéairement entre les deux valeurs qui l'encadrent.

Chaque station a reçu un poids provisoire d'après le type d'appareil et la fréquence des observations. Pour des instruments qui déterminent simultanément φ et TU0 (PZT et Astrolabes), le poids des mesures de TU0 a été généralement pris égal à la moitié du poids des mesures de φ.

Pour 1964 et 1965, l'heure définitive du BIH a été calculée par la méthode ancienne qui consistait à calculer d'abord les coordonnées du pôle, puis à faire une moyenne des TU2 demi-définitifs. Pour 1966, on a utilisé la solution d'après les Équations (1) et (2), mais comme les coordonnées du pôle avaient été calculées au préalable, la solution en TU2 a été corrigée pour qu'elle soit homogène avec ces coordonnées.

Avant d'adopter cette nouvelle méthode de calcul, nous l'avons essayée sur les années 1964 et 1965.

Nous donnerons maintenant quelques résultats et conclusions qui se dégagent de cette première étape de notre travail.

1. Valeur des mesures de temps

Il est apparu que la contribution des instruments des passages méridiens était très faible et nous avons établi séparément les polhodies déduites de ces instruments et des astrolabes et PZT (Figures 1 et 2).

La polhodie déduite des mesures de latitude seulement présente une meilleure homogénéité et elle a une forme sensiblement différente: la Figure 3 a été établie à partir des résultats des 25 meilleures séries, provenant d'instruments de tous types

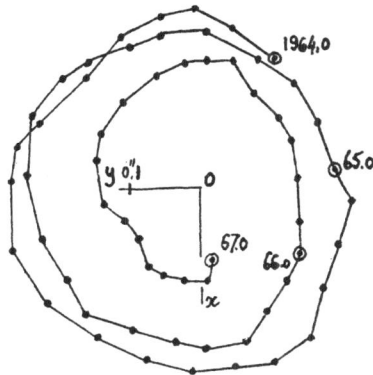

Fig. 3. *Polhodie d'après les mesures de latitude. 25 instruments divers, sans pondération.*

(dont la précision est comparable). Bien entendu la polhodie obtenue par la solution générale temps + latitude a une forme intermédiaire (Figure 4).

Il apparaît que le terme annuel de la polhodie a une forme différente suivant qu'il provient des mesures de temps ou de latitude (Figure 5). Notons que le terme annuel de la polhodie déduite des latitudes est comparable à celui obtenu par le SIMP. Il

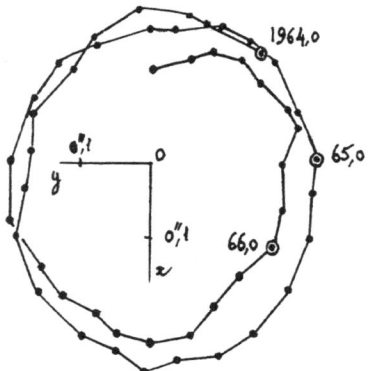

FIG. 4. *Polhodie d'après les mesures de temps et de latitude (PZT et astrolabes).*

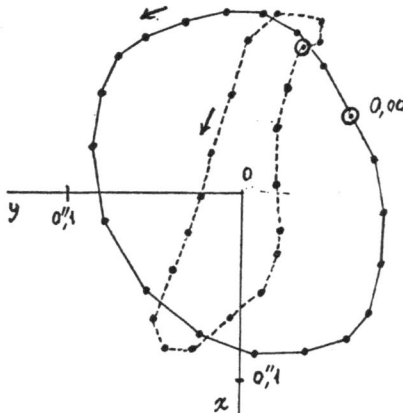

FIG. 5. *Polhodie annuelle; en trait plein, d'après les mesures de latitude; en pointillé, d'après les mesures de temps.*

convient de s'assurer que l'écart de la polhodie annuelle déduite des mesures de temps est bien réel. Pour cela nous sommes en train d'essayer divers groupements d'instruments et nous réduisons les principales séries de mesures de TU0 depuis 1960.

Les mesures de l'heure peuvent contribuer utilement au tracé de la polhodie, après pondération appropriée. Nous souhaitons donc que les calculs de TU1 par le BIH soient exécutés à partir de l'ensemble des mesures de temps et de latitude.

2. Pôle de référence

Les longitudes conventionnelles du BIH définissent un pôle de référence et la solution des Équations (2) donne les coordonnées du pôle par rapport à ce pôle de référence. Mais, par suite des décisions de l'UAI, le pôle utilisé pour passer de TU0 à TU1 est le "pôle moyen de l'époque".

Le choix du pôle moyen a été dû à la difficulté de conserver en chaque station une latitude de référence fixe par suite des erreurs instrumentales et personnelles et par suite des erreurs des catalogues d'étoiles. Nous allons estimer la dérive probable fictive du pôle due à ces erreurs en considérant les mesures de latitude faites dans la quarantaine de stations qui participent aux travaux du BIH.

A. ERREURS INSTRUMENTALES ET PERSONNELLES

Ces erreurs sont bornées. Elles affectent la précision de la polhodie mais elles ne peuvent pas simuler une dérive séculaire.

B. ERREURS SYSTÉMATIQUES SUR LES MOUVEMENTS PROPRES

Seule l'erreur du type $\Delta\mu'_\delta$ peut conduire à une dérive fictive du pôle. Nous déduisons des comparaisons entre le FK3, le FK4, le GC et le N30 que la dérive introduite en une station par l'erreur $\Delta\mu'_\delta$ de l'un de ces catalogues est probablement inférieure à $0''.003$ par an. De plus, les dérives des diverses stations ne sont pas indépendantes: les dérives sont identiques pour des stations à même latitude utilisant le même catalogue. Ainsi l'influence sur le pôle est très réduite. A titre d'exemple et pour fixer l'ordre de grandeur, nous avons supposé qu'il fallait appliquer les corrections FK4-GC à tous les instruments autres que les astrolabes (qui observent des étoiles du FK4). On trouve ainsi que la dérive fictive introduite par le GC aurait été pour les 40 stations utilisées: $0''.0004$ par an. La dérive doit être bien moindre si les catalogues sont au préalable ramenés au système du FK4.

C. ERREURS INDIVIDUELLES SUR LES MOUVEMENTS PROPRES

On caractérisera ces erreurs par leur écart type: $0''.005$ par an (pour la différence FK4-GC, on a environ $0''.004$ par an). Sur un programme contenant 100 étoiles, la dérive à craindre est $0''.0005$ par an. Pour l'ensemble des stations de latitude, ces erreurs ont un caractère accidentel et la dérive du pôle à craindre, pour les 40 stations actuelles est $0''.0001$ par an. Ces erreurs sont donc négligeables.

D. ERREURS DUES AUX RENOUVELLEMENTS DE PROGRAMMES

Nous supposerons qu'un programme contenant 100 étoiles est complètement changé, mais que les positions des étoiles restent prises dans le même catalogue. Si, par exemple, on prend les positions du GC, l'erreur accidentelle d'une déclinaison est de l'ordre de $0''.3$, le changement de programme introduira donc un saut probable de $0''.03$ dans la latitude. Si ce programme reste valable 10 ans, la dérive apparente probable est de l'ordre de $0''.003$ par an. Et si toutes les stations changent de programme tous les 10 ans, la dérive fictive probable du pôle est un peu inférieure à $0''.001$ par an.

Mais il faut noter que (1) certains instruments ne changent pas de programmes (astrolabes), (2) les renouvellements de programmes conservent le plus possible d'anciennes étoiles, (3) qu'à très long terme, cette cause d'erreur n'introduit pas de dérive fictive, car les erreurs sont bornées et aléatoires autour du système du catalogue.

Evaluons cependant la dérive probable due à des changements de programmes analogues à celui du SIMP en 1967,0 (40% des étoiles renouvelées, pour 12 ans). Compte tenu du nombre d'astrolabes, cette dérive probable est de $0''.0002$ par an.

Toutes les dérives fictives que nous trouvons sont extrêmement lentes (10–20 fois plus lentes que la dérive du pôle moyen par rapport au pôle de référence du SIMP).

Il n'y a plus aucune raison de conserver le "pôle moyen de l'époque". Nous souhaitons que les calculs du BIH soient exécutés à partir d'un système cohérent de latitudes et de longitudes fixes définissant un pôle de référence conventionnel.

Nous pensons même que, grace au nombre d'instruments en service, le pôle de référence pourrait être mieux conservé par cette méthode statistique, que par la méthode géométrique du SIL/SIMP, à cause des possibles mouvements relatifs des zéniths. L'usage simultané des mesures de temps, convenablement pondérées, apporterait encore une stabilité accrue au pôle de référence.

3. Précision des instruments

Après résolution des Équations (1) et (2) pour tous les observatoires, en 1964 et 1965, nous avons considéré les résidus R_1 et R_2 de chaque type d'équation, dans le sens observation-calcul. Ces résidus, outre des écarts accidentels, présentent généralement une variation périodique annuelle et une variation progressive due, pour une faible part, à une dérive réelle ou fictive et, pour le reste, à des variations d'erreurs systématiques instrumentales. Nous prendrons l'absence de cette variation progressive comme critère de qualité des instruments, pour l'étude des dérives continentales et de la dérive du pôle. Nous calculons la variation progressive par différence des résidus moyens R_1 et R_2 pour chaque année:

$$d_i = (R_i) \, 1964 - (R_i) \, 1965, \quad i = 1, 2.$$

Nous considérerons la moyenne quadratique σ_i des d_i par type d'instruments.
Le tableau 1 donne les valeurs de σ_1 et σ_2.
Les PZT sont les meilleurs des instruments, à la fois pour les mesures de temps et de

Tableau 1

Stabilité des instruments sur un intervalle d'un an: on donne la moyenne quadratique des dérives constatées entre 1964 et 1965

Type des instruments	Nombre des instruments	Latitude, σ_1 toutes stations	Latitude, σ_1 toutes stations sauf une, anormale	Temps, σ_2 toutes stations
Astrolabes	7	0″021	0″021	0˸0070
PZT	9	0″015	0″015	0˸0019
Lunettes zénithales des latitudes	18	0″050	0″035	———
Instruments des passages	22	———	———	0˸0140

latitude. Mais, pour les travaux courants du BIH, il convient de rattacher leurs catalogues au FK4.

Les astrolabes A. Danjon peuvent donner des résultats équivalents en qualité à ceux des PZT, lorsqu'ils sont utilisés dans les meilleures conditions. En moyenne les résultats sont moins bons. Ils sont, de plus, entachés d'erreurs personnelles. Cependant, les astrolabes ont l'avantage de permettre des observations d'étoiles fondamentales, sans renouvellements périodiques de programmes. Il conviendrait d'étudier d'autres types d'astrolabes, en particulier des astrolabes photoélectriques.

Les lunettes zénithales visuelles modernes (ZTL 180) ont une précision comparable à celle des astrolabes et PZT pour les mesures de latitude. Les instruments anciens sont un peu moins bons.

Les instruments des passages méridiens, visuels ou photoélectriques, donnent des mesures de l'heure entachées d'erreurs systématiques, importantes et lentement variables avec des amplitudes qui dépassent parfois 50 ms. Compte tenu de ces erreurs, on est conduit à leur donner un poids quasi-nul dans le calcul de la polhodie et de l'heure. Il ne faut pas oublier qu'à un rapport des erreurs de 10 qui est pratiquement le rapport

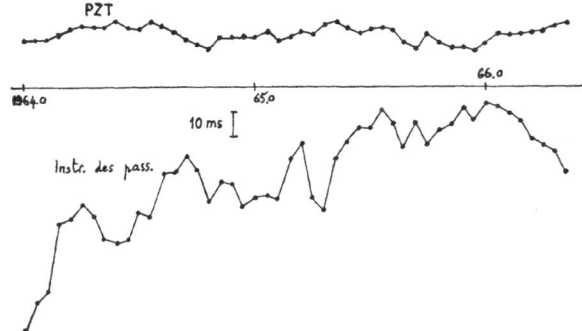

FIG. 6. *Exemple typique des erreurs des mesures du temps pour deux types d'instruments.*

entre les erreurs des instruments des passages et des PZT, il correspond un rapport des poids de 1/100. Voir Figure 6.

Les dérives continentales devraient être étudiées essentiellement par des PZT placés sur des parallèles communs et observant des listes d'étoiles communes. Les observations d'astrolabes sont rapportées à un catalogue suffisamment homogène, qu'elles contribuent de plus en plus à améliorer, pour qu'il soit plutôt recommandable de disperser les instruments afin d'étudier les dérives entre points de latitude différente (Europe–Amérique du Sud, par exemple).

Enfin on peut regretter que les stations du SIMP ne recueillent pas d'information sur l'heure.

Les stations du SIMP devraient être équipées de PZT.

Recommandations

Les parties du texte en italique constituent les recommandations que nous souhaiterions voir adopter.

4.2. CONTRIBUTION APPORTÉE PAR LES MARÉES TERRESTRES DANS L'ÉTUDE DE LA ROTATION DE LA TERRE

P. MELCHIOR

(Observatoire Royal de Belgique, Uccle, Bruxelles, Belgique)

RÉSUMÉ

Trois aspects de l'étude expérimentale des marées terrestres présentent un intérêt direct pour l'étude de la rotation de la Terre: (1) La détermination expérimentale des nombres de Love et l'étude des effects dynamiques du noyau liquide sur les nutations de l'axe principal d'inertie; (2) Le retard de la marée terrestre; et (3) Les dérives des pendules horizontaux.

ABSTRACT

Three aspects of the experimental study of earth-tides present a direct interest for the study of the rotation of the earth. These are (1) the experimental determination of the Love numbers and the dynamical effects of the liquid core on nutations of the principal axis of inertia; (2) the delay of earth-tides; and (3) the drift of horizontal pendulums.

1. Effets du noyau liquide sur les ondes tesserales diurnes et sur les nutations associées

Le problème théorique consiste à exprimer les nombres de Love h, k en fonction de la répartition interne des propriétés élastiques et de la densité.

En 1957, Jeffreys reprenant une étude antérieure de Poincaré démontre qu'un effet de résonance dû aux mouvements dans le noyau liquide se manifeste sur les ondes dont la période est suffisamment voisine de celle du jour sidéral (K_1 et, dans une moindre mesure P_1). Or ce sont précisément les composantes horizontales de ces ondes de marée qui créent les couples de précession et de nutation:

l'onde K_1 correspond à la précession et à la nutation de 18,6 an

l'onde O_1 correspond à la nutation semi-mensuelle

l'onde P_1 correspond à la nutation semi-annuelle.

L'amplitude théorique de ces ondes est respectivement de $0''.0070$, $0''.0050$, $0''.0023$ mais la précision d'une seule mesure horaire au pendule horizontal atteint $0''.0002$.

Les nutations associées ont des amplitudes théoriques respectives de $9''.22$, $0''.09$ et $0''.55$ tandis que la précision d'une observation au cercle méridien atteint au maximum $0''.2$. En outre la période semi-annuelle n'est pas favorable à une détermination astronomique expérimentale.

On voit ainsi l'intérêt évident qu'il y a à passer de la méthode de mesure d'angles,

Markowitz and Guinot (eds.), Continental Drift, 71–76. © I.A.U.

purement géométrique, des astronomes à la méthode dynamique consistant à mesurer directement à l'aide de pendules horizontaux les forces qui créent les nutations.

Jeffreys et Vicente (1957) ont traité une série de modèles à noyau liquide. Molodensky développe une étude analogue en 1961 et obtient des résultats assez semblables à ceux de Jeffreys et Vicente (Molodensky, 1961).

Selon ces calculs théoriques les trois ondes diurnes que l'on peut espérer déduire avec quelque précision des enregistrements auront pour coefficients d'amplitude les nombres reproduits dans le Tableau 1.

Tableau 1

	Déviations de la verticale facteur $\gamma = 1 + k - h$				Variations de g facteur $\delta = 1 + h - \frac{3}{2}k$			
	(1)	(2)	(3)	(4)	(1)	(2)	(3)	(4)
K_1	0·714	0·693	0·734	0·730	1·183	1·185	1·137	1·143
P_1	0·695	0·696	0·699	0·697	1·196	1·172	1·154	1·158
O_1	0·654	0·658	0·688	0·686	1·224	1·211	1·159	1·164
$(O-K_1)$	−0·060	−0·035	−0·046	−0·044	+0·041	+0·026	+0·022	+0·021

(1) Modèle de Jeffreys et Vicente.
(2) Modèle de Jeffreys et Vicente avec particule centrale.
(3) Modèle de Molodensky avec noyau fluide homogène.
(4) Modèle de Molodensky avec graine centrale.

L'effet de résonance atteint donc près de 10% de l'amplitude totale.

Une relation indiquée par Melchior (1950):

$$k/h = 1/\lambda = \left(\int_0^a \rho r^4 \, dr\right)\Bigg/\left(\int_0^a \rho r^2 \, dr\right) \tag{1}$$

soit encore

$$k/h = \left(e - \frac{q}{2}\right)\Bigg/[(C - A)/A] = 1 - (2/5)\sqrt{1 + \eta} = (3/2)(C/Ma^2) \sim 1/2,$$

où λ est le paramètre de d'Alembert et η le paramètre de Radau ($q = \omega^2 a/2g$, $e =$ aplatissement) est retrouvée par Molodensky (1961) mais non dans les modèles de Jeffreys et Vicente et c'est la raison des désaccords assez prononcés qui existent entre ces deux théories pour le facteur δ.

Nous avions trouvé en 1951:

$$k/h = 0·504$$

et G. Jobert avait démontré que si l'on fait abstraction de l'hypothèse implicite en (1) que les déformations sont homothétiques par rapport au centre, on peut seulement écrire (Jobert, 1952):

$$k/h \leq 0·504.$$

La discussion de résultat expérimentaux portant sur 19 414 jours d'observations en 27 stations fondamentales nous a conduits à la solution présentée dans le Tableau 2 (Melchior, 1966).

Tableau 2

Résultats expérimentaux

Ondes	γ	δ	k	h	k/h
K_1	0·747	1·143	0·220	0·473	0·465
P_1	0·721	1·148	0·262	0·541	0·484
O_1	0·676	1·160	0·328	0·652	*0·503*

Les effets dynamiques sont conformes aux théories bien qu'un peu plus prononcés que prévu pour le facteur γ.

La solution pratiquement statique qui correspond à l'onde O_1 appelle les remarques suivantes:

(1) elle donne pour le facteur δ une valeur égale à celle d'Asie Centrale;

(2) ce même facteur δ est égal à celui trouvé par Lecolazet et Steinmetz pour l'onde zonale $Mf (\delta = 1·16 \pm 0·09)$ qui obéit à la théorie statique;

(3) la valeur de k concorde avec celle trouvée par Markowitz dans les variations de la vitesse de rotation de la Terre résultant des variations d'aplatissement dues à la même onde $Mf (k = 0·34 \pm 0·07)$;

(4) elle restitue pratiquement la valeur 0·504 que nous avions trouvée pour le rapport k/h et s'accorde bien avec les modèles de l'intérieur de la Terre construits par Kaula;

(5) cette valeur de k donnerait cependant au mouvement du pôle une période (de Chandler) de 460 jours, ce qui est excessif et mérite examen. Il résulte de l'incertitude où nous sommes quant à la valeur exacte de cette période, que les procédés de moyennes mobiles utilisés pour calculer le mouvement séculaire du pôle doivent être remplacés par des méthodes de filtrage plus adéquates (filtres passe bande).

Ces résultats subiront certes encore quelques retouches lorsqu'on apportera les petites corrections dues aux effets indirects. Mais ceux-ci sont faibles pour les ondes diurnes et nous ne pensons pas que l'essentiel des conclusions puisse en être affecté.

2. Les ondes semi-diurnes et le freinage de la rotation de la Terre

Les marées terrestres de période semi-diurne sont plus difficiles à interpréter car les océans et en particulier les mers atlantiques côtières sont le siège d'importantes marées de caractère semi-diurne parce qu'elles appartiennent à des systèmes qui entrent en résonance dans cette gamme de fréquences.

Il s'ensuit que les appareils enregistreurs subissent sur ces fréquences d'importants effets perturbateurs que l'on appelle effets indirects parce qu'ils ont toutes les caractéristiques de la marée mais sont transmis par l'intermédiaire des marées océaniques.

Tableau 3

Composante Verticale. Déphasage de l'onde M_2

Atlantique		Centre		U.R.S.S.	
Bruxelles I	$+0°72$	Kieselbach	$-4°00$	Poulkovo	$-2°09$
Bruxelles II	$+1°56$	Potsdam	$-0°98$	Krasnaya P	$-4°23$
Dourbes	$+0°96$	Berggieshübel	$-2°98$	Moscou	$-1°34$
Vedrin	$+0°03$	Tihany	$-0°37$	Kiev	$-2°82$
Battice	$+0°78$	Borowiec	$-1°46$	Poltava	$-1°30$
Luxembourg	$+0°74$			Tbilissi	$(+1°20)?$
Strasbourg	$+2°08$	*Nord*			
Karlsruhe	$+0°34$	*Stockholm*	$-2°65$	*Asie*	
Frankfurt/M.	$+0°72$	*Helsinki*	$-0°41$	*Talgar*	$-3°43$
Bonn	$+0°89$			Tashkent	$-4°10$
Hannover	$+1°37$	*Sud*		Alma Ata	$-3°51$
Berlin	$+0°68$	Sofia	$-1°45$	Frounze	$-3°60$
Bad Salzungen	$+0°60$	*Genova*	$-0°20$	Langchow	$-2°88$
Freiberg	$+0°20$	*Trieste*	$-0°80$	*Téhéran*	$-3°67$
		Costozza	$-2°18$	*Kyoto*	$-2°17$
		Resina	$-2°72$		

$+$ = avance, $-$ = retard.

Les stations en italique sont celles pour lesquelles une analyse globale a été faite sur une très longue série d'observations (de 250 à 1000 jours).

Si l'on examine le Tableau 3, on constate que les effets indirects ne pourront guère justifier les différences de caractère régional que présente le groupement des stations. Ces effets indirects sont d'ailleurs très faibles pour l'onde K_1 qui présente cependant sensiblement la même répartition de phases que l'onde M_2.

Sans vouloir entrer ici dans les discussions de détail, nous pouvons résumer les faits comme suit:

Les déphasages sont faibles en Europe Occidentale et Centrale, aussi bien pour l'onde diurne K_1 que pour l'onde semi-diurne M_2. Ces mêmes déphasages sont par contre très prononcés pour K_1 aussi bien que pour M_2 en Russie, en Asie Centrale et au Japon.

Le Tableau 3 montre que l'on peut délimiter trois zones:

Europe Atlantique:	avance de $0°7$ environ
Europe Centrale et Italie:	retard de $1°7$ environ
URSS, Asie et Japon:	retard de $3°5$ environ

Or, ces phases concernent la marée globale.

Ce qui nous intéresse c'est le déphasage de la marée de déformation et celui-ci est six fois plus grand si le facteur d'amplitude δ est égal à 1·20.

Un retard de la déformation de type sectoriel (onde M_2) provoque évidemment un freinage de la vitesse de rotation de la Terre, car le couple exercé par les composantes Est–Ouest de la force de marée sur cette déformation radiale de marée n'est plus nul si le bourrelet est décalé par rapport à l'axe Terre–Lune.

On constate ici la présence de déformations relatives entre diverses régions de l'écorce terrestre ce qui donnerait lieu à une dissipation d'énergie additionnelle.

On peut interpréter les résultats repris dans la Tableau 3 en disant que le bourrelet de marée aborde le Japon et l'Asie avec un retard sensible qu'il rattrape rapidement au cours de sa progression à travers l'Europe, le tableau montre que le Japon est solidaire de l'Asie ce que confirment les mesures de longitude et de latitude présentées au cours de ce Symposium.

3. Contribution des stations clinométriques à l'étude des mouvements récents de l'écorce terrestre

Dans une note récente (Melchior et Brouet, 1966) nous avons montré que les pendules horizontaux en quartz, bien installés (c'est-à-dire directement dans la roche) présentent des dérives rapidement décroissantes liées à la procédure d'installation.

Après deux ou trois ans de fonctionnement cette dérive est de l'ordre de $0''.002$ par jour et l'appareil tend vers une stabilité presque totale.

Dès lors on peut espérer obtenir à longue échéance des informations de caractère nouveau si l'on dispose d'un réseau de stations assez nombreuses pour permettre l'élimination des effets purement régionaux.

Les perturbations locales les plus dangereuses sont de caractère hydrologique (crues de rivières et de nappes aquifères souterraines).

Une étude plus récente a été menée avec un système d'étalonnage automatique programmé réalisant la mesure de la sensibilité des appareils deux ou trois fois par semaine.

Deux appareils (VM76 et VM77) ainsi installés dans la mine de Přibram (Tchécoslovaquie) à une profondeur de 1500 m, ont une sensibilité qui présente une très faible variation rigoureusement linéaire :

$$s = 0''.0013 - 0''.0000017/\text{jour}$$

par millimètre à 5 m (focale) (7 mois d'observations avec 72 mesures automatiques de s à chaque appareil).

Il est souhaitable qu'un bon réseau de stations bien équipées soit développé sans tarder.

4. Conclusion

Nous estimons que les problèmes de la rotation de la Terre et de ses déformations ne peuvent plus être abordés séparément.

Bibliographie

Jeffreys, H., Vicente, R.O. (1957)   The Theory of Nutation and the Variation of Latitude,

Mon. Not. R. astr. Soc., **117**, 142–161. – The Theory of Nutation and the Variation of Latitude: The Roche model core, *Mon. Not. R. astr. Soc.*, **117**, 162–173.

Jobert, G. (1952)   Marées terrestres d'un globe fluide hétérogène, *Ann. géophys.*, **8**, 106–111.

Melchior, P. J. (1950)   Sur l'influence de la loi de répartition des densités à l'intérieur de la Terre dans les variations Luni-Solaires de la gravité en un point, *Geofis. pura appl.*, Milano, **XVI**, 105–112; ou *Commun. Obs. r. Belgique*, **195** (*Géoph.*, **20**).

Melchior, P. (1966)   *The Earth Tides*, Pergamon Press, New York, 458 pp.

Melchior, P., Brouet, J. (1966)   Contribution des stations clinométriques de marées terrestres à l'étude des mouvements récents de l'écorce, *IIe Symp. Intern. sur les Mouvements récents de l'Ecorce Terrestre, Helsinki 1965, Commun. Obs. r. Belgique*, **B8**, (*Géoph.*, **75**), 275–281.

Molodensky, M.S. (1961)   The Theory of Nutations and Diurnal Earth Tides, *IVe Symp. Intern. sur les Marées Terrestres, Commun. Obs. r. Belgique*, **188** (*Géoph.*, **58**), 25–56.

4.3. THE POLAR MOTION DERIVED FROM BOTH TIME AND LATITUDE OBSERVATIONS

SIGETUGU TAKAGI

(International Latitude Observatory, Mizusawa, Japan)

ABSTRACT

The coordinates of the pole were derived from both time and latitude observations made at eight observatories from 1962 to 1964. The differences between these results and the ILS results show an annual variation. After removal of this term the standard deviation is about $\pm 0''03$ for the x- and y-components. Corrections to the adopted longitudes and latitudes are obtained and these agree closely with those obtained by the BIH.

RÉSUMÉ

On a déduit les coordonnées du pôle simultanément des observations de temps et de latitude faites par huit observatoires de 1962 à 1964. La différence entre les résultats obtenus et ceux du SIL montre une variation annuelle. Après correction pour ce terme, la moyenne quadratique des différences est environ $0''03$ pour les coordonnées x et y. On obtient des corrections aux longitudes et latitudes adoptées qui concordent bien avec celles obtenues par le BIH.

1. We have attempted to derive the polar motion by making use of time and latitude observations simultaneously. We used the observations made with modern instruments, PZT and astrolabe, from 1962·0 to 1964·5 at eight observatories: Mizusawa, Tokyo, Paris, Greenwich, Alger, Neuchâtel, Washington, and Richmond.

2. The times observed are reduced to the atomic time system of A.1. The times observed at Greenwich and Paris are reduced to the A.1 system by the comparison between the standard clocks and the A.1 system, published in the Bulletins of the observatories. Those of other observatories are reduced to the A.1 system by using the results of the reception of the time signals.

3. The results of the observations made each year were divided into twelve groups by taking the monthly mean. The first nominal epoch of the monthly-mean date is 1962·0. The results were treated by methods of the modern theory of statistics. The weight of the results of every clear night is taken as unity. The weight is usually taken as the number of stars observed, but in the case where there are few observations the weight thus defined often makes skew the apparent distribution of data over the interval considered. The mean date of the monthly mean is made as near as possible to the nominal date within a range of ± 3 days, in order to avoid a

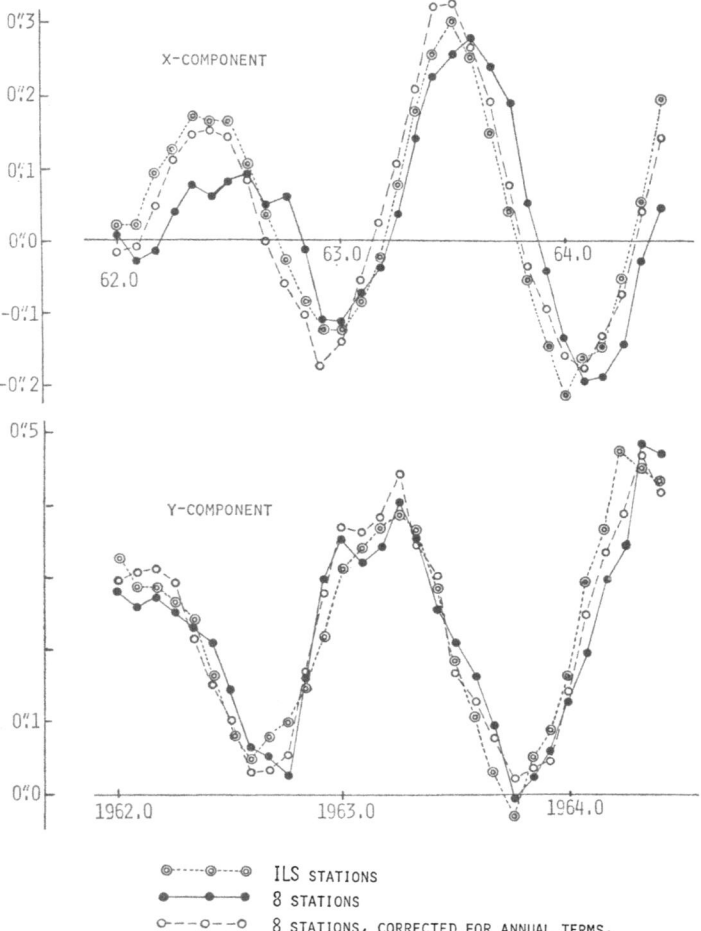

FIG. 1. *Coordinates of the pole derived from time and latitude observations and derived from ILS observations.*

timing error in the analysis. However, in some cases we had to obtain the data by means of interpolation.

4. The data obtained as above are combined to derive the coordinates of the pole by the following equations

$$\Delta\varphi = x \cos \lambda + y \sin \lambda + z$$
$$\Delta T = (x \sin \lambda - y \cos \lambda) \tan \varphi + s.$$

The coordinates are shown in Figure 1 together with those of the ILS. The difference between the coordinates in this paper and those of the ILS seems to be large and show an annual variation. After removal of the annual term from the differences, the standard deviations between the two coordinates are given as follows:

$$x\text{-component} \quad \pm 0\overset{''}{.}031$$
$$y\text{-component} \quad \pm \overset{''}{.}032.$$

The annual term may be chiefly due to a difference in the star catalogs used.

5. The coordinates of the mean pole in the period from 1962·0 to 1964·5 are determined with respect to both BIH and ILS system and are as follows.

	BIH system	ILS system
	$x = + \overset{''}{.}0684$	$x = + \overset{''}{.}0209$
	$y = + \overset{''}{.}0622$	$y = - \overset{''}{.}1593.$

The BIH mean pole with respect to the ILS system in the same period is

$$x = - \overset{''}{.}0475$$
$$y = - \overset{''}{.}2215.$$

The coordinates given in this paper are reduced to the BIH system. The mean pole given above is rather different from that of the BIH. This suggests that the time system and the latitude system adopted do not have the same reference system.

The corrections to the longitude and latitude of the observatories have been obtained above with respect to the BIH system. These are shown in Table 1, together with the corrections which are given annually in the Bulletins of the BIH. They are reduced to the conventional longitude system of 1962 and to the mean of the ten observatories which are not influenced by the polar motion.

It is interesting that in general the longitude values of this paper and those of the BIH (1962–63) agree well for the American and Asian stations. We cannot be sure of the variations in these values for a long interval. It is supposed that the difference in the y-component of the mean pole causes a discrepancy.

Table 1

Station	Latitude	Longitude This paper	Longitude BIH (1958–61)	Longitude BIH (1962–63)
Washington	$+ \overset{''}{.}0220$	$+ \overset{s}{.}0040$	$+ \overset{s}{.}0141$	$+ \overset{s}{.}0039$
Richmond	$+ \overset{''}{.}0857$	$+ \overset{s}{.}0068$	—	$+ \overset{s}{.}0083$
Tokyo	$+ \overset{''}{.}0193$	$- \overset{s}{.}0086$	$- \overset{s}{.}0010$	$- \overset{s}{.}0028$
Paris	$- \overset{''}{.}0216$	$+ \overset{s}{.}0038$	$+ \overset{s}{.}0043$	$- \overset{s}{.}0023$
Greenwich	$- \overset{''}{.}0091$	$+ \overset{s}{.}0107$	$+ \overset{s}{.}0044$	$+ \overset{s}{.}0009$
Mizusawa	$- \overset{''}{.}0173$	$- \overset{s}{.}0134$	$- \overset{s}{.}0041$	$- \overset{s}{.}0135$
Neuchâtel	$- \overset{''}{.}0224$	$- \overset{s}{.}0005$	$+ \overset{s}{.}0040$	$- \overset{s}{.}0065$
Alger	$- \overset{''}{.}1423$	$- \overset{s}{.}0021$	$- \overset{s}{.}0090$	$- \overset{s}{.}0147$

4.4. POLAR WANDERING AND CONTINENTAL DRIFT

S. K. Runcorn

(School of Physics, The University of Newcastle upon Tyne, England)

ABSTRACT

The relations between paleomagnetic observations, polar wandering curves, and changes in the speed of rotation of the Earth are discussed.

RÉSUMÉ

Les relations entre les observations paléomagnétiques, les trajectoires du pôle et leurs changements, la vitesse de la rotation de la Terre sont discutées.

1. The astronomer, studying the motion of the pole over the last hundred years, and the palaeomagnetist face the same problem: that of distinguishing between relative displacements of continental blocks and motions of the Earth (or at least the crust) as a whole with respect to the pole. Both are measuring essentially the same quantity: the variation of latitude of a place on the Earth's surface and essential to each is the assumption of a fixed axis. The astronomer is able for practical purposes to suppose that the axis of instantaneous rotation of the Earth, that is, the axis of the celestial sphere, remains fixed in space, for it is always within $0''.001$ of the axis of angular momentum. The precession and nutation of this axis is a separate question. The corresponding invariable axis in palaeomagnetism is that of the mean geomagnetic field, which, when averaged over times represented in a geological strata, tens or hundreds of thousands of years, is a dipole aligned along the axis of the Earth's rotation. The direction from which the astronomer measures the latitude of a place is the direction of the vertical or the vector g. Corresponding to this the palaeomagnetist can obtain from the remanent magnetization of rocks the mean direction of the Earth's magnetic field (**H**) for some geological period. In both measurements, then, the essential question is the relation between a vector at a point on a continent and the Earth's axis of rotation.

The power of the astronomer's tool is the extreme accuracy with which he may measure directions (a hundredth of a second of arc). Because of the secular variation of the geomagnetic field, arising from magnetohydrodynamic turbulence in the Earth's fluid, electrically conducting, core, the field direction at a place at the present time diverges and over historical and geological times on the average varies by $20°$ from that of a dipole parallel to the Earth's axis of rotation. The secular variation is bound to be averaged out when a rock formation is sampled and the mean of the remanent magnetization vectors of the rocks taken. But the palaeomagnetist cannot indefinitely improve the accuracy of **H** by taking larger numbers of specimens, for systematic sources of error are present in his determinations. Firstly, secondary magnetization may be present in the rock as well as the primary magnetization acquired from the geomagnetic field at the time of the rock's formation. It may be possible to eliminate this secondary magnetization by "magnetic washing" techniques in the laboratory or to allow for it by examining the spatial relation of the remanent magnetization of a folded rock strata but it is impossible to do this with an accuracy of much better than a degree. Secondly, the primary magnetization, although acquired from the ambient geomagnetic field, at the time of cooling in the case of a lava or during the deposition of a sediment, may not be exactly parallel to it because of some anisotropy in the rock. Thirdly, although the original horizontal in a rock is preserved, especially in the case of a bedding plane in a sediment, the plane is never exactly specified, so that an allowance for tilting of the rocks since they became magnetized cannot be done with extreme accuracy. Thus **H** cannot be determined more accurately than, say, a degree. The geological age of the rocks in question cannot usually be known more accurately than 1 m.y. But the powerfulness of this tool at the disposal of the geophysicist is the extreme length of time over which he can recover from the rocks the field direction; even Pre-Cambrian sediments and volcanics have stable remanent moments. His measurements are almost a million times less accurate than those of the astronomers but they extend over ten million times as long a period! The variations discovered in both fields are concerned with the internal mechanics of the Earth and each should throw light on research in the other field, as I hope to show.

It is well established that for one continent for one geological time the palaeomagnetic vectors are consistent with the mean geomagnetic field being a geocentric dipole. The axis or pole of this, however, shifts with geological time. Had contemporaneous poles for different continents, determined from remanent magnetism, been in agreement for all the geological periods, polar wandering alone, in which the crust (and possibly the mantle) had moved as a whole relative to the axis of rotation, would have been a sufficient explanation of the palaeomagnetic data. The mean palaeomagnetic field – like the present one – is often specified by the angle of declination (D) and the angle of dip or inclination (I). At a latitude (λ), the dipole formula gives

$$\tan I = 2 \tan \lambda.$$

Latitudes thus determined of continents in remote geological times are compared with the evidence of climates found in the geological record and within each continent are now widely thought to be in satisfactory accord.

It is still useful, however, to represent the palaeomagnetic results from one continent by a polar wandering curve and it is found that the overall motion is a few tenths to a few degrees per million years. Because age determinations in geology have accuracies ranging from a tenth of a million years in the Tertiary to a million years in the Palaeozoic to tens or even hundreds of millions of years in the Pre-Cambrian, time discrimination in the geological record is poor. Thus even if the longest period of the geomagnetic secular variation was only the 500 years, which we detect from historical observations of D and I, we could not hope palaeomagnetically to examine excursions of the pole of periods as short as a few thousand years. Palaeomagnetism therefore cannot show whether the pole moves continuously or in steps.

2. The first palaeomagnetic studies of the geological column in Great Britain suggested polar wandering to be occurring and this led at once to the test of making palaeomagnetic measurements in another part of the world, the Grand Canyon, as a means of examining whether polar wandering alone explained the phenomenon or whether relative continental displacements had to be assumed. Similarly in the last century the variation of latitude data obtained by European observatories led to similar sets of observations being made elsewhere and the Waikiki results especially established that the main phenomenon being observed was a nutation of the pole of figure about the pole of the Earth's rotation.

Divergence between the polar wandering paths from different continents is held to have established continental drift, a hypothesis first advanced by A. Wegener by qualitative geological reasoning. There is general agreement that the continents were formerly grouped in the two continents of Gondwanaland and Laurasia at the end of the Palaeozoic and that the dispersal of the continents to form the Atlantic and Indian Oceans and to close the Tethys Sea has taken place in the last 100–200 million years. Thus, relative motions of the continental blocks of a few thousand kilometers in a few hundred million years seem established. This overall rate of movement of 1 cm/year is of the same order of magnitude as movements observed geodetically in the last 50 years within continents along transcurrent faults and as the spreading of the ocean floor on either side of the oceanic ridges inferred from magnetic surveys. These measurements do not show the horizontal displacements of the continents to be discontinuous although during earthquakes local displacements of meters can occur.

While relative displacements of the continents are occurring, continental drift and polar wandering cannot be separated, but it may be that prior to the dispersal of Gondwanaland and Laurasia or in the earlier evolution of the earth, or, at intervals only, polar wandering occurred and the continents remained in the same relative positions in the crust or perhaps drifted orders of magnitude less swiftly than in the last 100 million years. These are questions which will be answered as the palaeomagnetism of the Palaeozoic and Pre-Cambrian becomes better known.

It appears to be established that the pole has moved 0″.220 in the last 60 years (see Markowitz, 1967). The question whether there are relative movements of the continents has not yet been resolved from the International Latitude Service data. The origin of this secular motion of the pole of $0''.004$/year will now be discussed. Polar wandering inferred from palaeomagnetic observations appears to take place as well as continental drift. No explanation of drift has been suggested which is as convincing as the hypothesis that below a rigid crust, there is a spherical shell – whether this is the whole of the mantle is under dispute – in which flow, of thousands of kilometers in scale, is occurring at a rate of a few cm to 1 m per year. This is possible for the Earth's solid interior must, because of the rapid rise of temperature with depth and because of the exponential dependence of creep processes on temperature, behave, over millions of years, as a fluid below about 50 km; although above this depth it acts as an elastic solid which fractures when stresses exceed certain limits. Such a hypothesis is also necessary to explain isostasy.

Convection currents (Runcorn, 1957) are capable of producing polar wandering. The polar wandering curves determined from Europe and North America give a mean rate since the Pre-Cambrian of $\frac{1}{3}°$ per million years or $0''.001$ per year (Creer *et al.*, 1957). This rate is not constant, however, and between the Devonian and the Upper Carboniferous, palaeomagnetic data from Australia (Irving, 1966) and South America (Creer, 1965) seems to show movements which may be about a factor of 5 faster.

It is not unreasonable therefore to suppose that the secular motion of the pole discovered from observations of the International Latitude Service is the same phenomena as that of palaeomagnetism. The explanation is therefore to be found in the flow pattern of the Earth's mantle. It has been argued that the recently determined low harmonics of the geopotential arise from density variations in the Earth's mantle associated with these flow patterns (Runcorn, 1966, 1967). Thus it appears that the present-day flow distribution may be worked out and as the geoid becomes better determined, it may be possible to calculate the expected secular motion of the pole. The variation of latitude data will then provide a test of the theory of convection currents in the Earth's mantle.

3. The geomagnetic secular variation has its origin in the Earth's fluid core; its time scale is 10–1000 years, much too short to be explained by physical processes in the Earth's mantle. Hydromagnetic theory applied to the Earth's core has perhaps had its most significant success in explaining the irregular fluctuations in the length of the day, the existence of which astronomers proved from the study of the motions of the Sun, Moon, Venus, and Mercury. It is significant that the irregular fluctuations in the length of the day, the geomagnetic secular variation and the unknown process which excites the Chandlerian nutation, are the only phenomena connected with the Earth's interior which have time scales of the order of 10–100 years.

It has often been suggested that earthquakes are the source of excitation of the Chandlerian nutation, as indeed they were once suggested as a cause of the irregular fluctuations in the length of the day – the major ones occurring on a similar time scale.

The latter is easily seen to be quantitatively wrong. Movements of mass associated with earthquakes are orders of magnitude too small to change the moment of inertia of the mantle by the few parts in 10^7 required to explain the irregular changes in the length of the day. The excitation required to cause the nutation is a jolt of an axis through $0\overset{''}{.}1$ of arc every ten or few tens of years or more frequent smaller ones. To suddenly move the axis of figure by this amount requires a change in the inertia tensor of 1 in 2×10^6. We can demonstrate that this is most unlikely.

Imagine the Earth to consist of a spherically symmetrical body, of moment of inertia I, and an external equatorial bulge of moment of inertia $I/300$. Suppose the two can move relative to one another as may happen in the crust during an earthquake. A torque $I\omega\dot{\theta}$ applied about an equatorial axis to the sphere, will cause it to rotate about the axis at right angles to this at a rate $\dot{\theta}$, where ω is the Earth's angular velocity. The equal and opposite torque on the equatorial bulge will cause it to rotate with an angular velocity $300\,\dot{\theta}$ about the same axis. Thus to move the pole by $0\overset{''}{.}1$ of arc, the crust exemplified here by the equatorial bulge, 20 km in height, would have to move 1 km. It is hard to imagine that this can be realistically supposed to happen in an earthquake, as Mansinha and Smylie (1967) have suggested.

We must therefore enquire if rather sudden changes of the direction of the axis of angular momentum of the Earth's mantle of the order of $0\overset{''}{.}1$ are possible. The Earth's core is known to be rotating at a present rate of $\frac{1}{5}°$ per year westwards relative to the mantle and it was pointed out by Runcorn (1954) that a change in this of 20% would imply a change in the length of the day of 1/300 sec – such as was observed around 1897 – so that the total angular momentum of the Earth may remain constant. To effect this transfer of angular momentum between the core and mantle, we must postulate the existence of an impulsive torque equal to 2×10^{-7} of the Earth's angular momentum. It is believed that this arises from the currents induced in the lower part of the Earth's mantle, which has an electrical conductivity of 0.1–10 ohm^{-1} cm^{-1} as expected from the semi-conducting properties of silicates and oxides. The theory of this electromagnetic coupling has not yet been fully worked out but it is a necessary consequence of the observed existence of the secular variation of the geomagnetic field.

The secular variation, and the consequent mantle-core torques, arise from magneto-hydrodynamic turbulence in the core. It is not simply described, having no preponderant harmonic term as has the main field. I conclude therefore that the torques exerted on the mantle are not axially symmetrical. Thus, while the component parallel to the axis of rotation changes the length of the day, that in the Earth's equatorial plane moves the axis of angular momentum away from the pole of figure. As the effects are within an order of magnitude, I conclude that this is the cause of the excitation of the Chandlerian nutation.

The astronomical data on the variation of the length of the day was once taken to imply sudden or instantaneous changes. This is physically impossible, yet one must be impressed by the short time constants involved. This seems to be possible as the time constants of free decay of current systems in the lower mantle are of the order of 1 year. Turbulent disturbances pass through the core as magnetohydrodynamic waves with velocities of 1–10 cm per second or over distances equal to the scale of the non-dipole field anomalies in months or years. This fact suggests a reason both for the comparative sharpness of the changes in the length of the day and of the jolts seen in the nutation data.

Rather sudden changes in the rate of rotation of the core have been found by the study of the westward drift of the non-dipole field since 1830. Conservation of the Earth's angular momentum also requires in the direction of the total angular momentum of the core (including that of its eddies). The geomagnetic secular variation data should allow this prediction of the above theory of the excitation of the Chandlerian wobble to be tested.

Thus, the data of the International Latitude Service is one important key in the study of the dynamic processes in the mantle and core.

References

Creer, K.M. (1965)   *Phil. Trans. R. Soc. Lond.,* **A 258**, 27–40.
Creer, K.M., Irving, E., Runcorn, S.K. (1957)   *Phil. Trans. R. Soc. Lond.,* **A 250**, 144.
Irving, E. (1966)   *J. geophys. Res.,* **71**, 6025.
Mansinha, L., Smylie, D.E. (1967)   *J. geophys. Res.,* **72**, 4731–4743.
Markowitz, W. (1967)   the present volume, p. 25.
Runcorn, S.K. (1954)   *Trans. Amer. Geophys. Un.,* **35**, 49.
Runcorn, S.K. (1957)   in *Gedenkboek F.A. Vening Meinesz,* Mouton & Co, The Hague, pp. 271–277.
Runcorn, S.K. (1966)   *Geol. Survey Canada, "The World Rift System",* Paper 66-14.
Runcorn, S.K. (1967)   *Geophys. J. R. astr. Soc.,* **14**, 375–384.

5.1. INFORMATION OBTAINABLE FROM LASER RANGE MEASUREMENTS TO A LUNAR CORNER REFLECTOR

C. O. ALLEY

(Dept. of Physics and Astronomy, University of Maryland, College Park, Md., U.S.A.)

and

P. L. BENDER

(Joint Institute for Laboratory Astrophysics, University of Colorado and National Bureau of Standards, Boulder, Colo., U.S.A.)

ABSTRACT

It has been proposed to the U.S. National Aeronautics and Space Administration that optical retro-reflector packages be placed on the lunar surface under either the Surveyor or Apollo Programs. Methods for measuring the range to the reflectors with an expected accuracy of 15 cm have been presented. The new technique is briefly discussed, and an analysis of the determination of geocentric longitude is given, indicating a potential uncertainty of 0.25×10^{-3} s of time.

RÉSUMÉ

Il a été proposé à l'Administration nationale américaine de l'aéronautique et de l'espace (NASA) de poser sur la surface de la Lune un réflecteur artificiel de lumière – cataphote – dans le cadre de l'un de ses programmes – "Surveyor" ou "Apollo". Diverses méthodes ont été présentées, assurant aux mesures de la distance aux réflecteurs une précision de 15 cm. La technique nouvelle est discutée succinctement, et une analyse est présentée de la méthode de détermination de la longitude géocentrique avec une précision de 0.25×10^{-3} s horaires.

1. Introduction

In this paper we wish to suggest that precise measurement of the range between stations on the Earth and a bench mark on the Moon consisting of an optical retro-reflector, made possible by recent progress in quantum electronics and space capabilities, offers a very useful technique for the study of the topics of this symposium. Increased knowledge of the center of mass motion of the Moon, forced physical librations, and the lunar figure, as well as information on a possible secular variation of the gravitational constant, would also be produced by a successful experiment continued over one or two decades.

A brief description of the method has been published (Alley *et al.*, 1965). A much more detailed analysis of the science and technology has been made in a formal proposal to carry out the experiment, submitted to the National Aeronautics and Space Administration of the United States by a group* consisting of the present authors (who are representing the group) and D. Brouwer (Yale University, now deceased), D. G. Currie (University of Maryland), R. H. Dicke (Princeton University), J. E. Faller (Wesleyan University), G. J. F. MacDonald (University of California at Los Angeles), H. H. Plotkin (Goddard Space Flight Center), S. K. Poultney (University of Maryland), H. Richard (Goddard Space Flight Center), D. T. Wilkinson (Princeton University), U. Van Wijk (University of Maryland, now deceased), and W. M. Kaula (University of California at Los Angeles). Much of this material will be published elsewhere.

There are indications in the Soviet literature of similar plans for the emplacement of a lunar optical corner reflector (Kokurin *et al.*, 1966a, 1966b, 1967).

2. Technique

It is well within the capabilities of existing ruby lasers to generate a pulse of radiation of duration 10^{-8} s, with an energy of 10 joules at 6943 Å, and an angular radius of 10^{-3} rad from a 2-cm aperture (≈ 100 times the diffraction limit). If this beam is recollimated to exit from a good quality 150 cm aperture, it will have an angular radius of about 2.″7 and would illuminate an area on the Moon a few kilometers in extent. The presence of a nearly diffraction limited optical-corner reflector near the center of the pattern (which will be blotchy because of index variations in the atmosphere) will send back a part of the pulse to the receiver without any appreciable spreading in time, as contrasted with the return from the curved lunar surface which will be spread to several microseconds. Also, the return will be many times that of the surface for a modest sized reflector. For a currently considered design, having a mass of several kilograms, the enhancement is two orders of magnitude for an albedo of 10%. Thus, on a single shot it will be possible to measure the round trip time of ≈ 2.5 s to 10^{-8} s, yielding a range accuracy of 1.5 m. By processing the returns from ≈ 100 shots, it seems possible to determine the range to 15 cm.

If the transmitting aperture is used also for receiving, there will be a reduction in the signal caused by velocity aberration. For the parameters given above, assuming an enhanced quantum efficiency of 10%, a signal of a few photo-electrons is expected. The number of photo-electrons produced by the background radiation passed by a 3-Å filter during the 10 nanosecond duration of the pulse will be much smaller, even for the sunlit condition on the Moon. Therefore the signal is sufficient for measurement when a range gate is used. A computer program to predict the range to the necessary accuracy at the epoch of firing, which needs to be measured to 0.1 millisecond,

* In June, 1967, NASA endorsed the emplacement of a retro-reflector on the Moon as an official Apollo lunar surface experiment for the earliest possible landing. Responsibility for the design to accomplish the scientific objectives was given to this group.

is required. With an average power of about 3 watts, corresponding to one shot every 3 s, one could achieve the desired range accuracy in about 5 min.

It is expected that continued progress in laser technology will lead to smaller beam divergences and to shorter pulse lengths than have been used in the above analysis. Ground stations with smaller apertures and less complex electronic processing equipment will then be possible.

It should be emphasized, as is explained in more detail below, that the ranging technique is independent of:

(1) angular errors caused by the atmospheric index of refraction,

(2) knowledge of the local vertical direction.

3. Range Method for Measuring Geocentric Longitudes

In this paper we will limit ourselves mainly to the determination of differences in geocentric longitude between stations. Let t_M be the time at which the distance from an observing station to the reflector is minimum. We assume that measurements U_1 and U_2 of the distance are made at times t_1 and t_2 which are respectively about 4 hours before and 4 hours after t_M (see Figure 1). For simplicity, the motion of the Moon is neglected. If the difference in the observed distances is small, we find the following formula for t_M:

$$t_M = \tfrac{1}{2}(t_1 + t_2) + \frac{U_1 - U_2}{2\omega R \sin \phi}.$$

Here ω is the rotation rate of the earth, R is the distance of the station from the axis of rotation, and 2ϕ is the angle of rotation between t_1 and t_2. For $R = 5 \times 10^6$ m and an uncertainty of 15 cm in $U_1 - U_2$, the resulting uncertainty in t_M would be 0.25 millisec. From similar measurements at two stations the difference in geocentric longitude would be obtained.

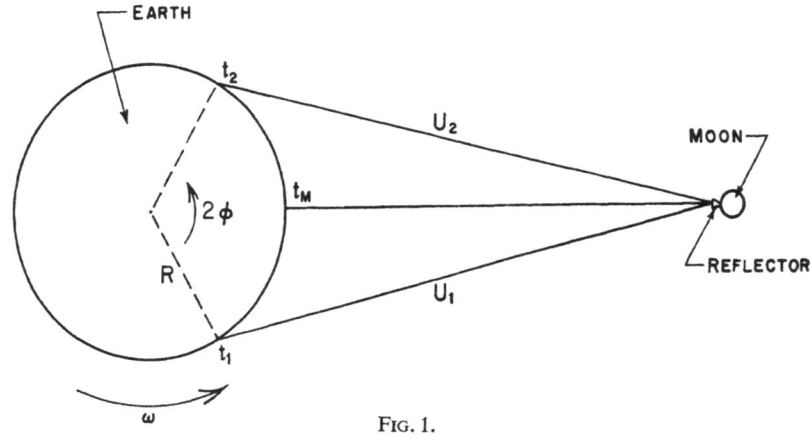

Fig. 1.

4. Accuracy of $U_1 - U_2$

Random Errors: For 10 nanosecond long laser pulses, about 100 pulses would be required to reduce the random error due to the pulse length to 15 cm. Since the use of a larger number of pulses per day appears practical and since information on longitude differences can be accumulated over extended periods, it is probable that the limitation will come from systematic errors.

Time Delays: We have estimated that the absolute calibration of time delays in the photo-multiplier and electronics can be done to somewhat less than one nanosecond, but in this application it is the difference in the time delays between the measurement of U_1 and U_2 which is important. While variation in the time delay over the eight hour interval between the measurements will probably be observed, there is no apparent reason for a systematic difference.

Earth Tides: The contribution from the earth tides would be expected to be nearly the same for the two measurements if the station is not too near the ocean.

Lunar Motion: After six months to a year of lunar distance measurements, it will be possible to obtain a good fit for the most important parameters in the Hill-Brown-Eckert theory of the lunar motion and for the lunar librations and location of the reflector with respect to the lunar center of mass. This will make it possible to predict the change over any period of eight hours in the distance from the center of the Earth to the reflector on the Moon with an accuracy of about 2 cm.

Atmospheric Index of Refraction: Range uncertainties due to this source have been studied extensively for microwaves (Bean and Thayer, 1963). At an elevation angle of 20 deg above the horizon, bending of the path contributes negligible error and the total extra optical path length due to the atmosphere is about 6 m. For light this additional path can be predicted to about 6 cm, which is a factor 2 better than for microwaves because of the much lower sensitivity to water vapor (G. D. Thayer, private communication). The difference between the atmospheric errors in U_1 and U_2 will not necessarily be smaller because of diurnal effects and time variations, but it should still be substantially reduced when the observed difference in longitude between two stations is averaged over a month.

It is worthwhile to consider the effect of an east–west gradient in the index of refraction across the station of 1×10^{-8} per km. This corresponds to a temperature gradient of $0 \cdot 01$ °C/km if the pressure is constant. We assume for simplicity that the atmosphere is uniform in density up to 10 km and zero above that height. For vertical observations, the above gradient leads to a deflection of $0 \cdot 02$ (of arc), or about a 60 cm error in the station position observed with a photographic zenith tube. The corresponding difference in the distances measured through the atmosphere at angles of $+60°$ and $-60°$ from the zenith due to this gradient is $0 \cdot 2$ cm. While the two distance measurements are actually made at times eight hours apart, the above example still gives reason to expect the mean value of the atmospheric error over extended periods to be much less for methods based on distance measurements than for those based on angle measurements.

5. Continental Drift and Polar Motion

The above discussion indicates that the systematic and random errors expected in determining the difference in geocentric longitude of two stations may be small. However, observations from a number of stations would be necessary in order to separate the effects of polar motion from changes in longitude due to crustal movements. While the cost of the necessary equipment is high, it seems desirable for the countries which can do so to undertake lunar range measurements if a suitable retroreflector is placed on the Moon.

6. Other Information

For the rotation of the Earth with respect to the Moon, some of the above arguments concerning the reduction of errors by averaging do not apply. An estimated accuracy of somewhat better than one millisec of time seems reasonable. It appears that the motion of the Moon with respect to the Sun would be determined with sufficient accuracy so that effectively the rotation of the Earth with respect to the Sun would be obtained. Changes in the difference in latitude of two stations could be measured with an accuracy of about 0.″01 (of arc). The distance of the observing station from the axis of rotation would be determined to about 30 cm and the distance above the equatorial plane to about 2 m. For stations at nearly the same latitude but about 120° different longitude, the distance between the stations would be determined to about 50 cm (relative longitude to about 15 cm).

References

Alley, C. O., Bender, P. L., Dicke, R. H., Faller, J. E., Franken, P. A., Plotkin, H. J., Wilkinson, D. T. (1965)   Optical Radar Using a Corner Reflector on the Moon, *J. geophys. Res.*, **70**, 2267.
Bean, B. R., Thayer, G. D. (1963)   Comparison of Observed Atmospheric Radar Refraction Effects with Values Predicted Through the Use of Surface Weather Observation, *J. Res. NBS*, **67D**, 273.
Kokurin, Yu. L., Kurbasov, V. V., Lobanov, V. F., Mozhzherin, V. M., Sukhanovskiy, A. N., Chernyk, N. S. (1966a)   Measurement of the Distance to the Moon by Optical Radar, *JETP Letters*, **3**, 139.
Kokurin, Yu. L., Kurbasov, V. V., Lobanov, V. F., Mozhzherin, V. M., Sukhanovskiy, A. N., Chernyk, N. S. (1966b)   On the Possibility of Measuring the Shape and Orbit Parameters of the Moon by Optical Location, *Kosmicheskiye Issledovaniya*, **4**, 414–426.
Kokurin, Yu. L., Kurbasov, V. V., Lobanov, V. F., Mozhzherin, V. M., Sukhanovskiy, A. N., Chernyk, N. S. (1967)   On the Possibility of Refining by Means of Optical Location Some Astronomical Parameters of the System Earth–Moon, *Kosmicheskiye Issledovaniya*, **5**, 219–224.

5.2. INTERCONTINENTAL LONGITUDE DIFFERENCES OF TRACKING STATIONS AS DETERMINED FROM RADIO-TRACKING DATA*

DONALD W. TRASK and CHARLES J. VEGOS

(Jet Propulsion Laboratory, Pasadena, Calif., U.S.A.)

ABSTRACT

Radio tracking of artificial satellites has provided longitude differences between stations on different continents to 10 m. An accuracy of 1 m in the future is expected.

RÉSUMÉ

L'observation par radio des satellites artificiels a fourni des différences de longitude entre stations sur des continents différents à 10 m près. On espère une précision de 1 m dans le futur.

1. Introduction

The difference between *absolute* longitudes of two tracking stations located on different continents has already been determined to less than 10 m while an accuracy on the order of 1 m is expected in the future.

The determination of these station locations as well as certain physical constants such as the masses of the Earth and Moon, are a by-product of the precise radio tracking of deep space missions performed by the Deep Space Instrumentation Facility (DSIF). The relationship between the knowledge of the spacecraft trajectory obtained from this tracking data and the deep-space station (DSS) location solutions along with the current work and associated limitations are discussed in this paper.

2. Discussion

The DSIF provides a precise measurement of the change in slant range ($\Delta\rho$) between the spacecraft and the tracking station over a given time interval (T_c). Currently $\sigma_{\Delta\rho} < 10$ cm is obtainable for $T_c \leq 10$ min at lunar distances. It has been shown by Hamilton and Melbourne (1966) that the information content of a single pass (horizon to horizon of a single DSS) of Doppler for a distant spacecraft can be compressed

---

* This paper presents the results of one phase of research carried out at the Jet Propulsion Laboratory, California Institute of Technology, under Contract No. NAS-7-100, sponsored by the National Aeronautics and Space Administration.

into three parameters (a, b, c); namely,

$$\left.\begin{aligned}
a &= \dot{r} \\
b &= \omega r_s \cos \delta \\
c &= (\alpha - \lambda - \omega \xi)
\end{aligned}\right\} \tag{1}$$

where $\dot{r}$ = probe geocentric radial velocity, ω = Earth spin rate, δ = probe declination, α = probe right ascension ($\alpha_\odot \sim$ instantaneous right ascension of the mean Sun, assumed well known compared to the other parameters), t_s = DSS time reference converted to universal time, ξ = error in converting the time reference at the DSS to true universal time, $\Delta \dot{\rho}_s = r_s \omega \cos \delta \sin(\omega t_s + \alpha - \alpha_\odot - \lambda - \omega \xi)$ = range rate due to tracking-station motion, and $\dot{\rho} = \dot{r} + \Delta \dot{\rho}_s$ = range rate of probe relative to DSS.

The Doppler tracking data are primarily sensitive to the distance* of the tracking station off the Earth's spin axis r_s and the longitude λ of the tracking station. Changes in the station location parallel to the Earth's spin axis have little effect on the Doppler data; i.e., the tracking station is essentially located on the surface of a cylinder by the tracking data. Notice that b of equation (1) represents the amplitude of $\Delta \dot{\rho}_s$ while c is a measure of the time of meridian passage. It can be seen from equation (1) that station-location errors ($\varepsilon_{r_s}, \varepsilon_\lambda$) are highly correlated with probe angular position errors ($\varepsilon_\alpha, \varepsilon_\delta$), and that timing errors directly affect the solution for λ by an amount $\omega \xi$. In general, r_s and longitude differences are the better determined DSS coordinates. Longitude differences are not affected by timing errors common to the DSS or by errors in α. The tracking data directly determines the parameters a, b, and c; and hence the strength of these determinations is directly related to the quality of the tracking data. However, even if these parameters are perfectly known, the ability to determine tracking-station locations is still dependent upon one's knowledge of the spacecraft coordinates. That is, space missions which result in the probe being significantly influenced by the gravitational field of the Moon (and thus fixing the position of the spacecraft in space), as in the case of the Ranger lunar missions, will produce the best determinations of absolute tracking-station locations.

To date the most intensive postflight analysis has been performed on the four Ranger Block III (Rangers VI–IX) lunar missions. All four flights were tracked continuously from injection to impact on the lunar surface by the use of three DSS; namely, Goldstone, Calif., U.S.A. (DSS 12), Woomera, Australia (DSS 41), and Johannesburg, South Africa (DSS 51). This work was reported in Sjogren *et al.* (1966) and Vegos and Trask (1967a, 1967b), and current progress is regularly reported in Volume III of the JPL Space Program Summaries, issued bimonthly.

---

* For convenience, this report quotes distances in meters, although the true unit is the light meter (i.e., the distance the electromagnetic signal travels in a unit time is the basic unit of length in the "radio-tracking world"). Throughout the JPL analysis, lengths (in meters) depend on the adopted value of $c = 299\,792 \cdot 5$ km/s.

During the normal postflight analysis of the tracking data for each Ranger mission, a "best fit" to the tracking data was obtained by adjusting the probe coordinates at an epoch, mass of the Earth, mass of the Moon, and the solar-radiation effect on the spacecraft as well as the tracking-station locations. The longitude differences determined from each flight are compared to a statistical combination of the four flights in Table 1. The maximum deviation from the average for any of the flights is 10 m. The

Table 1

Longitude Differences Referenced to Pole of 1903·0

Source	Longitude Differences		
	$\lambda_{12} - \lambda_{41}$	$\lambda_{41} - \lambda_{51}$	$\lambda_{51} - \lambda_{12}$
Combined Rangers, deg	106·30701	109·20209	−215·50910
Ranger VI minus Comb. Rangers	4 m	− 7 m	3 m
Ranger VII minus Comb. Rangers	− 2 m	7 m	− 5 m
Ranger VIII minus Comb. Rangers	0	− 5 m	5 m
Ranger IX minus Comb. Rangers	− 10 m	8	2

statistical combination was accomplished in a least-squares sense by constraining those parameters common to each flight to a single best value. An exception to this was the Earth–Moon ephemeris scale factor which is allowed to have a unique value for each flight. This is done to compensate for errors which may exist in the lunar ephemeris. The actual statistics of the combined Ranger analysis are probably on the order of $\sigma_{\lambda_i - \lambda_j} \sim 10$ m. In addition to the quality of the tracking data and the uncertainties accounted for is the "fitters model" of the real universe; this statistic includes uncertainties in the Earth–Moon ephemeris, timing relationships, polar motion, and the effects of the ionosphere. Of these effects, the ionosphere is probably the dominant error source that has yet to be modeled and may account for errors from 5 to 10 m depending upon the particular flight. It is anticipated that further refinements in the postflight processing of existing tracking data will produce DSS longitude differences approaching 1 m. After the Ranger Block III missions (1965) the DSIF was converted from L-band to S-band which is only $\frac{1}{6}$ as sensitive to the ionosphere. The use of this S-band data may allow these 1-m goals to be exceeded. The navigational accuracy goals being considered for future missions (HAMILTON et al., 1967) will require the knowledge of *absolute* locations on the order of 1 m.

References

Hamilton, T.W., Melbourne, W.G. (1966)   Information Content of a Single Pass of Doppler Data From a Distant Spacecraft, *Space Programs Summary No. 37-39*, Vol III, Jet Propulsion Laboratory, Pasadena, Calif.

Hamilton, T. W., Grimes, D. C., Trask, D. W. (1967)   Critical Parameters in Determining the

Navigational Accuracy for a Deep Space Probe During the Planetary Encounter Phase, *Space Programs Summary No. 37-44*, Vol. III, Jet Propulsion Laboratory, Pasadena, Calif.

Sjogren, W. L., Trask, D. W., Vegos, C.J., Wollenhaupt, W. R. (1966)   *Physical Constants as Determined From Radio Tracking of the Ranger Lunar Probes*, Technical Report 32-1057, Jet Propulsion Laboratory, Pasadena, Calif.

Vegos, C. J., Trask, D. W. (1967*a*)   Tracking Station Locations as Determined by Radio Tracking Data: Comparison of Results Obtained From Combined Ranger Block III Missions and From Baker-Nunn Optical Data, *Space Programs Summary No. 37-43*, Vol. III, Jet Propulsion Laboratory, Pasadena, Calif.

Vegos, C. J., Trask, D. W. (1967*b*)   Ranger Combined Analysis, Part II: Determination of the Masses of the Earth and Moon From Radio Tracking Data, *Space Programs Summary No. 37-44*, Vol. III, Jet Propulsion Laboratory, Pasadena, Calif.

6.1. SUR L'UTILISATION DES DIFFÉRENCES DE LONGITUDE POUR LA DÉTERMINATION DU PÔLE INSTANTANÉ

A. Gougenheim

(Paris, France)

RÉSUMÉ

Pour déterminer la position du pôle instantané, on peut recourir, en dehors des méthodes de calcul, à une méthode graphique que l'on décrit.

ABSTRACT

To determine the position of the instantaneous pole one may use, in addition to calculations, a graphical method, which is described.

1. La méthode des moindres carrés, utilisée d'une manière générale dans le cas d'observations surabondantes exige que l'on rende d'abord linéaires les équations d'observation à l'aide d'une solution approchée. S'il n'y a que deux inconnues, ce qui est le cas des déterminations de position, les équations rendues linéaires peuvent être représentées sur un graphique à grande échelle par des droites qui sont les tangentes aux lieux géométriques correspondant aux diverses observations et dont le point de concours fournit la solution du problème. En général les droites ne concourent pas exactement et présentent une certaine dispersion due, partie aux erreurs d'observation, partie à l'incertitude affectant les éléments de base. Si l'on est gêné par cette dispersion pour adopter la solution, on peut toujours recourir à la méthode des moindres carrés, mais le graphique reste utile par la vue synthétique qu'il donne des observations et par l'approximation qu'il laisse prévoir pour le résultat.

2. Dès 1960, M. André Danjon avait montré qu'un tel graphique pouvait être établi pour la détermination du pôle instantané à l'aide de mesures de latitude. Une droite de latitude est perpendiculaire au méridien de la station d'observation et passe à une distance du pôle de référence, pris comme point approché, égale à la différence $\Delta\varphi$ entre la latitude mesurée et la latitude conventionnelle de la station.

3. Mais on peut également utiliser de la même façon les mesures de différence de longitude. Le lieu géométrique correspondant à une mesure est un segment capable sphérique décrit sur la base formée par les zéniths des deux stations en cause. Les

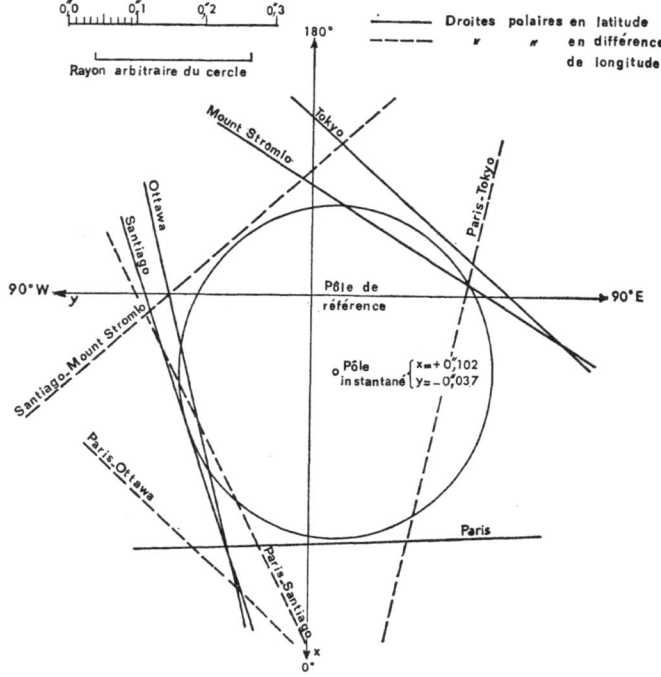

Fig. 1.

propriétés du segment capable sphérique montrent que le méridien du vertex de la base fait avec le méridien d'une des stations un angle égal à celui que le méridien de l'autre station fait avec le méridien normal au lieu, d'où la direction de celui-ci. En outre, si l'on appelle ΔG le petit écart entre la différence de longitude mesurée et la valeur conventionnelle qui correspond au pôle de référence, le lieu géométrique passe à une distance de ce dernier point égale au produit de ΔG par la cosécante de la différence de longitude et par la cotangente de la latitude du vertex.

La dispersion des divers lieux obtenus provient des erreurs d'observation, des erreurs des catalogues d'étoiles, et aussi, mais pour une très faible part, de la variation éventuelle des longitudes conventionnelles sous diverses actions, parmi lesquelles peut-être une dérive des continents.

4. Si les systèmes de latitudes conventionnelles et de longitudes conventionnelles correspondent exactement à un même pôle de référence, on peut faire concourir les droites de latitude et les droites de différence de longitude à la détermination du pôle instantané.

C'est dans cet esprit qu'a été établi l'exemple figurant sur la planche jointe, à l'aide de valeurs des coordonnées conventionnelles et des éléments d'observation de cinq stations; ces valeurs qui m'ont aimablement été communiquées par M. Bernard Guinot, directeur du Bureau International de l'Heure, se rapportent à une époque récente. Le graphique comporte cinq droites de latitude et quatre de différence de longitude. Sept de ces neuf lieux géométriques passent à l'intérieur d'une circonférence de 0″.03 de rayon. Aussi pour faciliter la résolution tous les lieux géométriques ont été déplacés parallèlement à eux-mêmes d'une quantité arbitraire (0″.225) de sorte que la résolution graphique consiste à déterminer le centre d'une circonférence de 0″.225 de rayon et tangentant au mieux l'ensemble des lieux géométriques.

6.2. FLUCTUATIONS IN THE MOTION OF THE MEAN POLE AND THE ROTATION OF THE EARTH

H. J. ABRAHAM

(Mount Stromlo and Siding Spring Observatory,
Australian National University)

ABSTRACT

Explanations of the progressive and librational motions of the pole are attempted.

RÉSUMÉ

On essaye une explication du mouvement progressif et périodique du pôle.

The observed secular motion of the mean pole shows a progressive component and periodic librations, either real or apparent. Evidence is given that real librations should occur.

The total deformation excitation Ψ due to yielding of the Earth should be proportional to kd, where k is the Love number and d is the distance of the pole of rotation from the pole of figure. Furthermore, the change in τ, the free nutation period, if due only to yielding of the Earth, should be a simple function of k. However, τ varies with d and therefore Ψ may include a function of d^2. The value of d is given by the free nutation term $m_0 \exp(i\sigma_0 t)$ plus the forced nutation terms $m \exp(i\sigma t)$ and $n \exp(-i\sigma t)$. Then it can be shown that these lead to mean excitations which vary as $\cos(\sigma_0 \pm \sigma) t/2$, and that one of these produces an excitation that varies as $\cos(\sigma - \sigma_0) t/4$. The frequencies $(\sigma - \sigma_0)/2$ and $(\sigma - \sigma_0)/4$ correspond to the observed libration periods of about 12 and 24 years.

The calculated meridian of the librations is in satisfactory agreement with the observed meridian, λ_L. The angle between λ_L and λ_p, the meridian of the progressive motion, is given by $\lambda = \cot^{-1}(k/k_f \cot\theta)$, where k_f is the value of k for hydrostatic equilibrium, and θ is the angle between λ_p and the major axis of the seasonal excitation.

The amplitude of the librations, and apparent fluctuations in the progressive motion, both vary with σ_0 in a manner to be expected. When the natural frequency σ_0 approaches the forced frequency σ the excitations are enhanced, amplitudes become greater, and the mean pole is advanced in the direction of the progressive motion.

Changes in T, the length of the day, show an effective value of k apparently varying with d. The observed value of T and the amplitude of the polar motion since 1955 show similar '6-yearly' variations proportional to $|\cos(\sigma-\sigma_0)\,t/2|$. The range of these variations in T relative to atomic time was 3×10^{-9}.

The effective value of k (and hence of τ) depends inversely on the mean excitation. When the librational excitation grows larger the values of τ grow smaller than otherwise. Moreover λ then grows larger, as it should because effectively k grows smaller. The '6-yearly' excitation which is caused by the resultant of the free and forced nutations shows a similar effect; when the resultant is greatest the excitations which it causes are greatest and the values of τ are least.

These relationships point to periodic excitations that connect the nutations with the librations and with changes in T.

6.3. THE DETECTION OF CHANGES IN THE COORDINATES OF A PLACE ON THE EARTH

R. O. VICENTE

(Faculty of Sciences, Lisbon, Portugal)

ABSTRACT

The accuracies involved in determining the motion of the pole and continental drift are discussed.

RÉSUMÉ

On discute les précisions requises pour déterminer le mouvement du pôle et la dérive des continents.

This paper computes the highest precision that can be attained by present-day methods for the observations of latitude, longitude and azimuth of a place. It is pointed out that there are only a restricted number of places on the Earth, at the present time, which can contribute to the detection of motions at the Earth's surface, and that most of these places are probably not situated in the regions of the Earth best suited for the researches concerned with motions in the upper layers of the Earth.

It is concluded that the highest precision, obtained by astronomical observations made during one night, of the coordinates of a point on the surface of the Earth gives a standard deviation of the order of 2·5 m in latitude and time determination. These values can be slightly improved by extending the period of observations. The precision obtained in observations of an astronomical azimuth is about 3 times lower. Attention is called to a certain number of systematic errors whose influence on the observations is not yet very well known.

It is stated that to observe motions of the ground of an order of magnitude of 0·25 m per year, and always in the same direction, we shall need at least 20 years and, even so, the standard deviation of the observations would be 50% of the displacement.

These values suggest that checking of possible motions of points on the surface of the Earth by astronomical observations is a very difficult task.

7.1. THE POLHODY IN A CRITICAL PERIOD
(1941·06–1948·98)

T. Nicolini

(Osservatorio di Capodimonte, Naples, Italy)

ABSTRACT

The definitive values of the coordinates of the pole from 1941·06 to 1948·98 as determined from ILS observations are given.

RÉSUMÉ

On donne les valeurs définitives des coordonnées du pôle de 1941·06 à 1948·98, déduites des observations du SIL.

The last-printed definitive results on the International polhody are those of the late Prof. L. Carnera, Director of the Central Bureau, for 1935·06–1940·98 (Carnera, 1957). Prof. Carnera also wished to complete the work from 1941·06 to 1948·98, the most critical period of the Service because of the second world war. He took the work to Florence, during his retirement, after the preliminary reductions had been made at the Naples Observatory. Several reasons, however, hindered the work, his age, the lack of help, the amount of numerical calculations, and above all, the decidedly deceiving preliminary results obtained. Also, confining the computations to the three stations responsible for the best observations, Mizusawa, Gaithersburg, and Ukiah, which had been in continuous operation during the whole period, caused difficulties. These problems led Prof. Carnera to drop the later calculations, of which he left no trace, in the declared conviction that nothing could be safely deduced.

However, a positive or negative conclusion must be given, and so, through the interest of the IAU, the IAG, and the Italian Geodetic Commission, I had the opportunity to resume the calculations already abandoned. By lucky chance Carnera, besides the preliminary results (given in several *Contributi* of Capodimonte (Carnera, 1947–9)) had published data useful for improved solutions (Carnera, 1953), including definitive values of the micrometer screws, temperature coefficients, level values, and corrections to the declinations. We must remember that because of the common programme observed by the stations on the same parallel, errors of declinations (and relative p.m.) give a common error in absolute latitudes, which is not our problem, and which is of no relevance to the polhody. The corrections to star declinations found by Carnera, could only give a better agreement between mean latitudes deduced from different groups, an improvement that could be more or less masked by the lack of precision.

Markowitz and Guinot (eds.), Continental Drift, 101–104. © *I.A.U.*

All the details for the final solution will be given in vol. X of Publications of the ILS. Here we shall only recall that more than one station has given impaired observations, and this would appear natural in the circumstances imposed by a world war. Unfortunately, Carloforte was also closed at times, beginning in 1941, and the difficulties of keeping skilled observers gave absurd results. The reopening took place in 1946.

After various trials, I found that the only combinations which gave consistent results are the following:

$$1941 \cdot 06 – 1946 \cdot 39, \text{ from M, K, G, U};$$

$$1946 \cdot 48 – 1948 \cdot 98, \text{ from M, K, C, G, U}.$$

About 50000 sheets of provisional calculations were retaken and corrected one by one, and mean latitudes for each stellar group were computed again. From these, equations for x, y and z, evening and morning, were established:

Table 1

Coordinates x and y of the North Pole 1941·060–1948·976

(Units: 0".001)

	1941		1942		1943		1944	
	x	y	x	y	x	y	x	y
·060	+066	+065	+156	+062	+089	+207	−011	+184
·143	+045	+037	+098	+074	+086	+195	+003	+229
·226	+022	+090	+075	+045	+132	+177	+025	+222
·310	−015	+101	−003	+113	+108	+134	+087	+228
·393	+008	+134	+035	+100	+153	+140	+194	+203
·476	+029	+177	+009	+110	+135	+082	+285	+171
·560	+069	+196	+023	+113	+122	+046	+302	+059
·643	+172	+209	+015	+112	+118	+043	+292	−008
·726	+137	+192	−005	+171	+040	+026	+161	−083
·810	+129	+154	−019	+170	−030	+050	+002	−088
·893	+148	+097	+033	+172	−085	+081	−059	−056
·976	+144	+083	+025	+225	−028	+091	−133	+012

	1945		1946		1947		1948	
	x	y	x	y	x	y	x	y
·060	−162	+098	−121	−057	+103	−075	+264	+108
·143	−158	+131	−189	+002	+007	−077	+237	+007
·226	−113	+238	−184	+086	−076	−028	+191	−019
·310	−053	+301	−145	+203	−150	+091	+106	−062
·393	+094	+310	−074	+285	−154	+160	+017	−023
·476	+200	+311	+062	+330	−084	+246	−032	+027
·560	+266	+172	+174	+346	+010	+285	−029	+067
·643	+286	+105	+219	+324	+119	+324	−038	+139
·726	+329	+010	+254	+268	+189	+311	−078	+184
·810	+252	−095	+309	+122	+263	+272	−032	+253
·893	+114	−137	+323	−030	+286	+227	+026	+305
·976	+064	−104	+233	−040	+278	+148	+129	+324

MKCGU,

$$x = -0·4359\varphi_M + 0·1227\varphi_K + 0·4484\varphi_C + 0·1233\varphi_G - 0·2583\varphi_U - 1·1694$$
$$y = -0·2637\varphi_M - 0·3133\varphi_K - 0·0172\varphi_C + 0·3382\varphi_G + 0·2559\varphi_U - 5·8770$$
$$z = +0·2305\varphi_M + 0·2007\varphi_K + 0·1756\varphi_C + 0·1851\varphi_G + 0·2082\varphi_U - 7·7337$$

MKGU,

$$x = -0·6288\varphi_M + 0·5580\varphi_K + 0·4291\varphi_G - 0·3584\varphi_U - 0·0971$$
$$y = -0·2563\varphi_M - 0·3300\varphi_K + 0·3265\varphi_G + 0·2597\varphi_U - 5·9181$$
$$z = +0·1550\varphi_M + 0·3712\varphi_K + 0·3048\varphi_G + 0·1690\varphi_U - 7·3132$$

Let A_i and A_i' denote, respectively, the coefficients of φ_i (mean monthly group latitude of station i) in the expression for x, and let a_i and a_i' be the analogous coefficients for

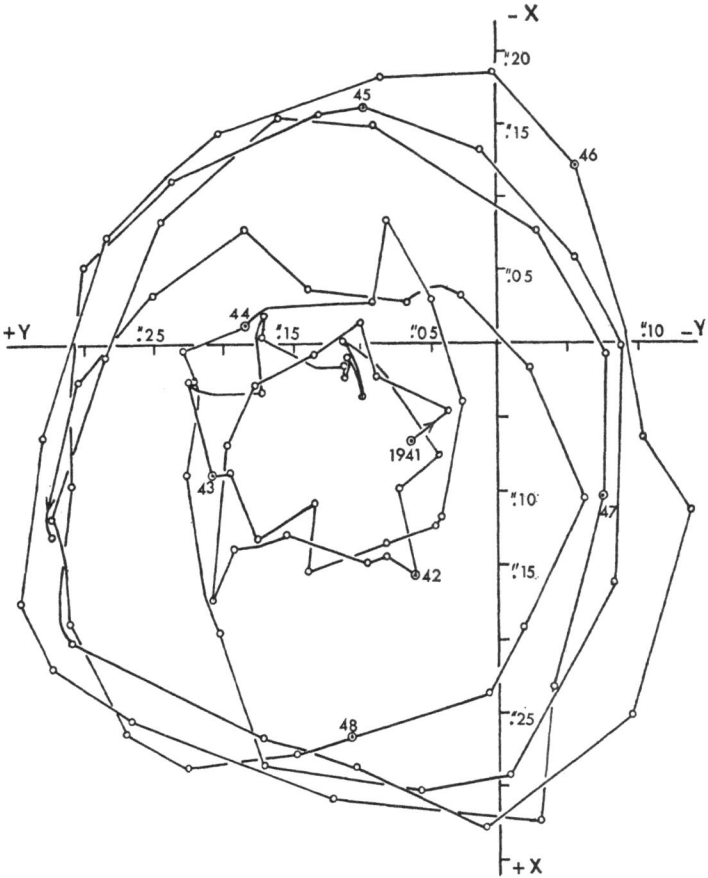

Fig. 1. *Polhody, 1941·06 to 1948·98.*

y. The conditions

$$\Sigma A_i = \Sigma a_i = \Sigma A'_i = \Sigma a'_i = 0.$$

are easily verified. If there is an error in the declination, the identity of the programme gives a common error in latitude, and the contributions of this error on *x*, or on *y*, are

$$\delta\varphi\Sigma A_i = \delta\varphi\Sigma a_i = \delta\varphi\Sigma A'_i = \delta\varphi\Sigma a'_i = 0.$$

This shows the independence of the polhody from errors in declinations.

I must add that different groupings of equations (that is different numbers of considered stations) give largely different solutions. It would be difficult to give the reasons for the peculiarities in the derived trend of the polhody. The solution from three stations, Mi, Ga, Uk, which I communicated for the Draft Reports of the 13th General Assembly of the IAU (Prague), is much less satisfactory than the one here reached.

The polar path resulting from my final computations are given in Table 1 and Figure 1.

References

Carnera, L. (1947–9)   *Contributi Astronomici di Capodimonte*, Serie II, **4**, Nos. 1, 2, 6.
Carnera, L. (1953)   Le livelle ed i micrometri dei telescopi zenitali nelle Stazioni Internazionali di Latitudine, *Memorie dell'Acc. Naz. dei Lincei*, 1953.
Carnera, L. (1957)   *Risultati del Servizio Internazionale delle Latitudini, dal 1935·0 al 1941·0*, vol. IX.

7.2. A PROPOSAL FOR A REFLECTING ASTROLABE

D. V. Thomas*

(Royal Observatory, Cape, South Africa)

ABSTRACT

A proposal is made for the construction of a 25 cm aperture "reflecting astrolabe", using a CER-VIT prism of near-zero thermal expansion. Such an instrument could be expected to have an advantage over the Danjon astrolabe as regards accuracy, declination-range and limiting magnitude.

RÉSUMÉ

On propose de construire un "astrolabe à réflexion" de 25 cm d'ouverture, en utilisant un prisme de CER-VIT à coefficient de dilatation quasi nul. Un tel instrument serait supérieur a l'astrolabe Danjon en ce qui concerne la précision, la zône de déclinaisons accessible et la magnitude limite.

Although the quality of the results obtained with Danjon astrolabes is generally high, the instrument does suffer from some defects and limitations. The angle of the main prism has been found to be very sensitive to temperature gradients induced in the prism by changes in the ambient temperature. In general there is a systematic increase in prism angle as the instrument cools during a night's observing. The result is a large "closing error" in prism angle, making reliable determination of group corrections in prism angle impossible. The residual zenith distance of an individual star also depends critically on the rank of the star in the group. Stars observable with an individual astrolabe are limited to a band less than 60° wide centred on the zenith. Thus the polar regions of the sky, particularly in the south, are not easily accessible. The limiting magnitude is about 6·0 in normal conditions, but the Paris (Guinot *et al.*, 1961) and Herstmonceux (Thomas and Wallis, 1967) results show that the observations of stars within one magnitude of the limit become increasingly affected by colour and magnitude equations, which are not the same for all observers. These defects affect particularly the performance of the instrument for the determination of catalogue corrections, but they also introduce difficulties in the precise comparison of the time and latitude results obtained by observation of the same stars at different sites.

The recent development of the glass-ceramic material CER-VIT offers the possibility of overcoming most of the above difficulties. The coefficient of expansion of

---

* Now at Royal Greenwich Observatory, Hailsham, Sussex, England.

Markowitz and Guinot (eds.), Continental Drift, 105–107. © *I.A.U.*

CER-VIT is more than a hundred times less than that of glass. Due to its crystalline structure, the material is generally unsuitable for transmission optics, but an aluminised CER-VIT prism, used as in Figure 1, could be expected to provide an altitude of observation stable to about $\pm\,.''01$.

A CER-VIT prism is being obtained for an experimental modification of the Danjon astrolabe at the Royal Observatory, Cape. A preliminary design study has also been undertaken of a "reflecting astrolabe" (with CER-VIT prism) having a total aperture of 25 cm, compared with 10 cm for the Danjon astrolabe (Thomas, 1967). The provisional layout of the optical components is illustrated in Figure 1, which may be compared with the corresponding diagram of the Danjon astrolabe (Danjon, 1958). The CER-VIT prism (A) has an angle of 135°, for observing at an altitude of 45°. It could be replaced by one with a smaller angle, for observing at a greater altitude, without much difficulty. The $f/10$ objective (C) has a focal length of 2·5 m. The quadruple Wollaston prism (D) has the same angle as the one in the Danjon astrolabe, as the focal ratios are the same, but rather more than double the linear

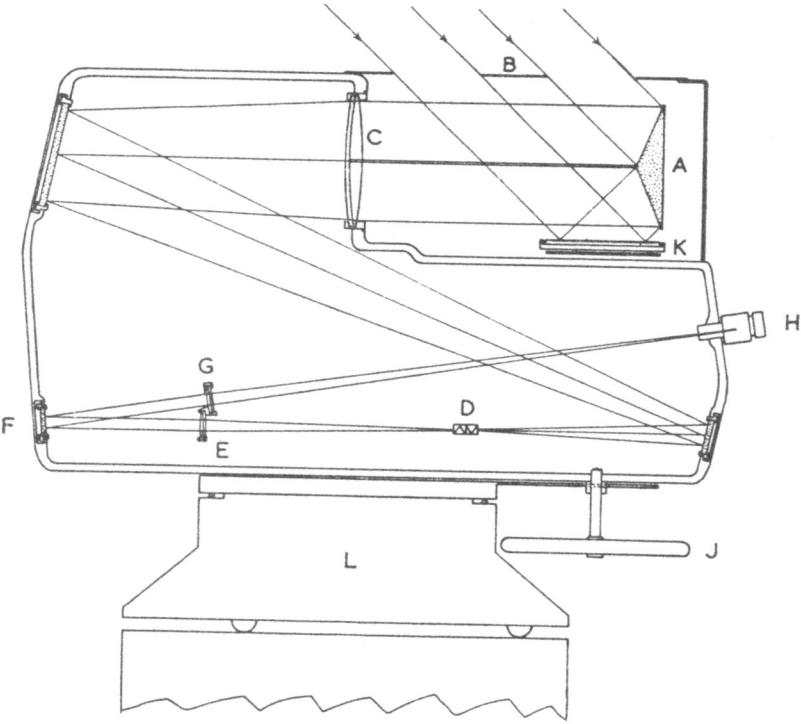

Fig. 1.   *Proposed 25 cm Reflecting Astrolabe with CER-VIT Prism. A, CER-VIT prism; B, removable wind-shield; C, 25 cm f/10 objective; D, Wollaston prism; E, G, transfer lenses; F, plane mirror; H, orthoscopic eyepiece; J, "steering wheel"; K, mercury bath; L, base (schematic).*

dimensions. The run of the Wollaston prism allows the change of altitude of a star to be compensated for a minimum of ± 20 s from the time of almucantar transit. The scale value of the micrometer is approximately 2″ per millimetre run of the Wollaston prism. A nominal reading accuracy correct to 0.″01 over the total range could easily be achieved by a commercial Moiré-fringe device using relatively coarse gratings, giving a digitised output direct on to punched cards or paper tape. The lens (G) has a focal length twice that of (E). The unwanted duplicated images formed by the Wollaston prism are removed by a stop placed immediately in front of (G), at the position of the image of the objective formed by (E). The 2·5 cm focal length orthoscopic eyepiece (H) gives a final magnification of 200, and a field of view 12′ in diameter. It is proposed that the "steering wheel" (J) should enable a differential speed variation to be imparted direct to the carriage of the Wollaston prism. This would avoid the difficulty of a given angular displacement of the images in the field requiring an angular rotation of the steering wheel proportional to the cosecant of the azimuth of observation. The provision of an automatic slow-motion device in azimuth would enable the stellar images to be kept stationary in the middle of the field during the observation, and help to eliminate personal equations depending on the azimuth of the star observed.

The detailed mechanical layout of the instrument has not yet been considered, but its construction should not present any insuperable difficulties.

In good conditions the proposed reflecting astrolabe could be expected to be more accurate than the Danjon astrolabe, due to greater stability of the altitude of observation, the slightly higher magnification (200 as against 175), and the smaller Airy disk (principal axes of first dark refraction ring 2.″5 × 1.″5). It could operate at any selected altitude above 45°, enabling both north and south poles to be reached from existing stations. The limiting magnitude of about 8·0 would bring all the stars of the FK4 and the majority of the stars of the GC within range. The brighter minor planets would be observable for varying periods from any station, providing a valuable link between the astrolabe system of stellar coordinates and the dynamical theory of the solar system.

References

Danjon, A. (1958)   *M.N.R.A.S.*, **118**, 411.
Guinot, B., Débarbat, S., Krieger-Fiel, J. (1961)   *Bull. astr.*, **23**, 307.
Thomas, D.V. (1967)   *Mon. Notes astr. Soc. Sth. Afr.*, **26**, 2.
Thomas, D.V., Wallis, R.E. (1967)   *R. Obs. Bull.* in preparation.

AUTHOR AND SUBJECT INDEX